Fragile Earth

Fragile

Earth

Views of a changing world

Fragile Earth
Views of a changing world

Collins
An imprint of HarperCollins*Publishers*
77–85 Fulham Palace Road
London
W6 8JB

First Published 2006

Printed by Imago in Thailand

British Library Cataloguing in Publication Data.
A catalogue record for this book is available from the British Library.

ISBN-13 978-0-00-723314-4
ISBN-10 0-00-723314-0

Front cover photograph:
Sossusvlei, Namib Naukluft Park, Namibia
© Lee Frost / Imagestate

Back cover photographs:
Muir Glacier, 1941 © NSIDC/William O. Field
Muir Glacier, 2004 © NSIDC/Bruce F. Molina

All mapping in this book is generated from Collins Bartholomew
digital databases. Collins Bartholomew, the UK's leading independent
geographical information supplier, can provide a digital, custom,
and premium mapping service to a variety of markets.
For further information:
Tel: +44 (0) 141 306 3752
e-mail: collinsbartholomew@harpercollins.co.uk

or visit our website at: **www.collinsbartholomew.com**

Collins. Do More.
www.collins.co.uk/fragileearth

Collins

Contents

Foreword

Sir Ranulph Fiennes, OBE

Back in 1982, with Charles Burton, I became the first person to reach both the north and south poles by surface travel. On the day we reached the North Pole, it seemed like another world which would be untouched by man forever. As we are all increasingly becoming aware, this is not the case. This superb book demonstrates that even remote wilderness areas within national parks are rapidly changing in their character. It presents a stark look at the sometimes catastrophic effects nature and mankind can have on the world.

As an explorer, one of the main challenges I face is the power and unpredictable character of the natural world. Although humans have come to dominate life on Earth, it is still the natural world which causes one of the biggest threats to civilization. The first chapter illustrates the power of the Earth and the constant changes resulting from major natural phenomena such as earthquakes and volcanoes. These events have been happening for thousands of years and have formed and changed landscapes around the world. The devastating effects of such events were seen in the Indian Ocean tsunami in December 2004.

It is not just natural events on or below the Earth's surface which threaten life and landscapes. Extreme weather has been a constant reminder of how vulnerable we are. The inhabitants of New Orleans discovered this to their cost in August 2005 when hurricane Katrina struck the coast of Louisiana. Seen through images from space, from the air and at ground level, the large-scale damage and change such wild weather can inflict becomes even more evident.

Natural disasters cause immediate change to a landscape, but man can cause equally dramatic changes over longer periods of time. As the world's population has increased, so has pressure on land. The images of Hong Kong show a striking example of how humans have completely changed a landscape. Massive building and development projects have expanded the city steadily upwards, and outwards into the sea.

11 September 2001 saw the famous Manhattan skyline in New York change forever. Human conflict, caused by many factors including ideology and competition over resources, can cause great damage to our world. Today's society is capable of impacting the world in irreversible ways and such changes can have devastating effects on people's lives and on the environment.

The world is now approximately 0.6°C (1.1°F) warmer than 100 years ago and there is clear evidence, as seen in the photographs presented, that many glaciers are now retreating and that the polar ice caps are shrinking. The rate of change appears to be increasing in both the northern and southern hemispheres with the effects likely to be felt around the world. The icy world of glaciers may feel a long way from our civilized world, but their shrinking could have a huge impact globally.

On my regular participation in adventure races, often in extremely dry climates, one of the most important factors is drinking enough water. It is water which helps to give you the energy for running the next mile. It is also what gives the Earth life and yet is increasingly lacking in some areas of the world. The scale of the shrinking of the Aral Sea in such a short period of time demonstrates how quickly an area can become desolate. In contrast, as land is developed and as the climate changes, increasing numbers of people are being put under threat of flooding from the sea, rivers and lakes. The sea also exerts a constant force on coastlines and can cause dramatic changes as material is washed away and deposited elsewhere.

Climate change and global warming were once the campaign words of environmental organizations. Now they have become the words of key scientific and government reports on the future prospects of the world. What does the future hold for our fragile Earth? Fascinating personal views from expert authors explore this question and look ahead to what the world might look like in the future.

The images in this book highlight dramatic changes which have happened to the world. They present a unique perspective on how the Earth is changing, and the new challenges we face. *Fragile Earth* is a wake up call. Is the world changing in the way that we want it to, or is it now out of our control?

Image locations

Locations of images found in the book,
and pages on which they first appear.

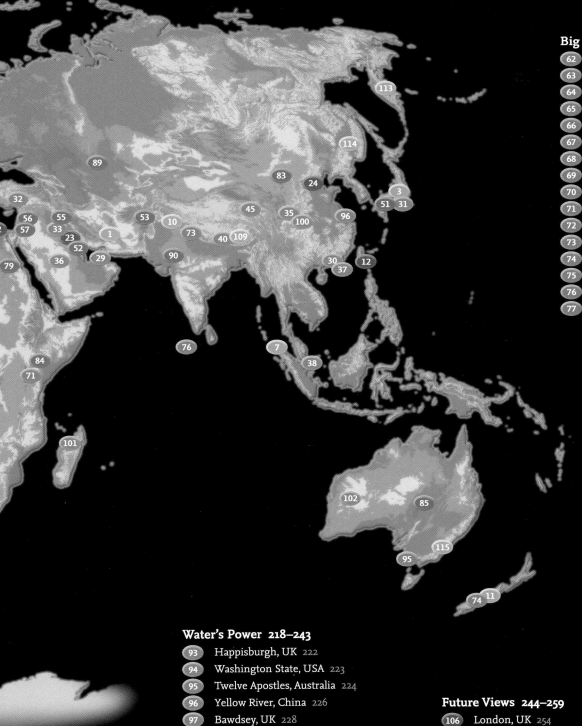

Restless

Earth

Natural phenomena

Earthquakes – *destructive movements of the Earth's surface due to sudden releases of energy in the Earth's crust or upper mantle*

Volcanoes – *openings in the Earth's crust from which molten lava, rock fragments, ash, dust, and gases are ejected*

Tsunamis – *large, often destructive, sea waves produced by submarine earthquakes, subsidence, or volcanic eruptions*

Landslides and avalanches – *movements of large masses of rock, snow or ice down mountains or cliffs*

The Earth's outermost solid layer consists of a thin crust of rock. Dense rock, approximately 5 km (3 miles) thick, lies beneath the ocean basins. Lighter rock forms the continents, projecting above the ocean surface. Where there are mountain chains the continental crust is up to 60 km (37 miles) thick. Although the crust covers the entire planet, it is not in the form of a single, unbroken skin. Crustal rocks are in sections of varying size, called plates. Individual plates can move, as though jostling with their neighbours, and the boundaries where plates meet are the locations of seismic activity – earthquakes and volcanic eruptions.

Deadliest earthquakes, 1900–2005

Year	Location	Deaths	Magnitude
1905	**Kangra**, India	19 000	7.5
1907	west of **Dushanbe**, Tajikistan	12 000	8.1
1908	**Messina**, Italy	110 000	7.2
1915	**Abruzzo**, Italy	35 000	7.5
1917	**Bali**, Indonesia	15 000	-
1920	**Ningxia Province**, China	200 000	7.8
1923	**Tōkyō**, Japan	142 807	7.9
1927	**Qinghai Province**, China	200 000	7.9
1932	**Gansu Province**, China	70 000	7.6
1933	**Sichuan Province**, China	10 000	7.4
1934	**Nepal/India**	10 700	8.1
1935	**Quetta**, Pakistan	30 000	7.5
1939	**Chillán**, Chile	28 000	8.3
1939	**Erzincan**, Turkey	32 700	7.8
1948	**Ashgabat**, Turkmenistan	19 800	7.3
1962	**Northwest Iran**	12 225	7.3
1970	**Huánuco Province**, Peru	66 794	7.9
1974	**Yunnan** and **Sichuan Provinces**, China	20 000	6.8
1975	**Liaoning Province**, China	10 000	7.0
1976	central **Guatemala**	22 778	7.5
1976	**Tangshan**, Hebei Province, China	255 000	7.5
1978	**Khorāsan Province**, Iran	20 000	7.8
1980	**Ech Chélif**, Algeria	11 000	7.7
1988	**Spitak**, Armenia	25 000	6.8
1990	**Manjil**, Iran	50 000	7.7
1999	**Kocaeli (İzmit)**, Turkey	17 000	7.6
2001	**Gujarat**, India	20 000	7.7
2003	**Bam**, Iran	26 271	6.6
2004	off **Sumatra**, Indian Ocean	>250 000	9.0
2005	**Pakistan**	80 000	7.6

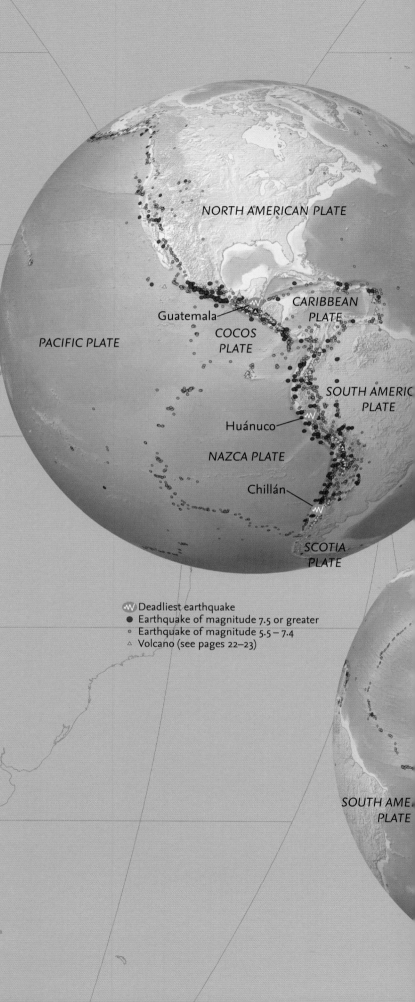

- Deadliest earthquake
- Earthquake of magnitude 7.5 or greater
- Earthquake of magnitude 5.5 – 7.4
- Volcano (see pages 22–23)

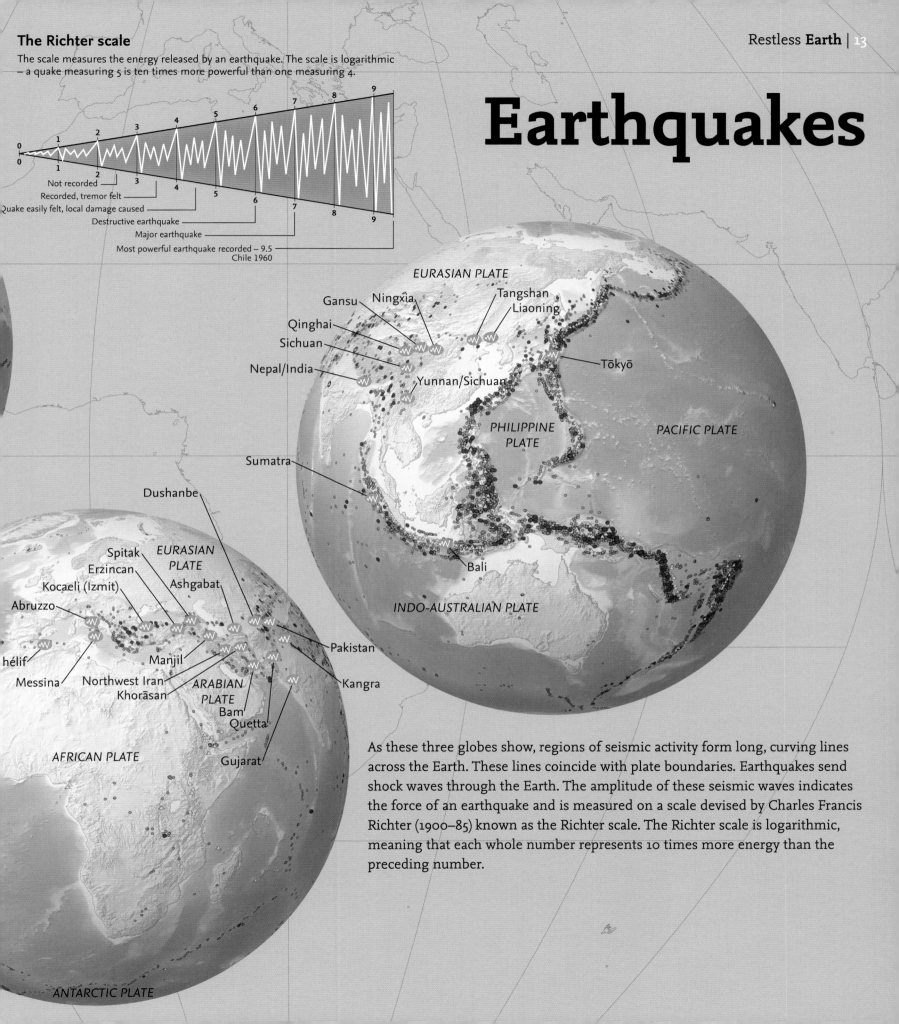

The Richter scale

The scale measures the energy released by an earthquake. The scale is logarithmic – a quake measuring 5 is ten times more powerful than one measuring 4.

Not recorded
Recorded, tremor felt
Quake easily felt, local damage caused
Destructive earthquake
Major earthquake
Most powerful earthquake recorded – 9.5
Chile 1960

Earthquakes

EURASIAN PLATE

Gansu
Ningxia
Qinghai
Sichuan
Tangshan
Liaoning
Tōkyō
Nepal/India
Yunnan/Sichuan
PHILIPPINE PLATE
PACIFIC PLATE
Sumatra

Dushanbe
Spitak
Erzincan
EURASIAN PLATE
Kocaeli (Izmit)
Ashgabat
Abruzzo
Chélif
Messina
Manjil
Northwest Iran
Khorāsan
ARABIAN PLATE
Bam
Quetta
Gujarat
Pakistan
Kangra
Bali
INDO-AUSTRALIAN PLATE
AFRICAN PLATE
ANTARCTIC PLATE

As these three globes show, regions of seismic activity form long, curving lines across the Earth. These lines coincide with plate boundaries. Earthquakes send shock waves through the Earth. The amplitude of these seismic waves indicates the force of an earthquake and is measured on a scale devised by Charles Francis Richter (1900–85) known as the Richter scale. The Richter scale is logarithmic, meaning that each whole number represents 10 times more energy than the preceding number.

The Earth's crust is divided into eight major plates: the African, Eurasian, Indo-Australian, Pacific, North American, South American, Nazca, and Antarctic. There are also minor plates, such as the Cocos, Caribbean, Philippine, Scotia and Arabian as well as even smaller microplates and fragments that remain from former plates which have disappeared. Arrows at plate boundaries on the map opposite indicate the direction in which the plates are moving. Rocks on one side of a plate boundary are jammed tightly against those on the opposite side. When the plates move they do so suddenly, with a jerk that relieves some or all of the stress along the join. Those jerks move the Earth's surface, causing the Earth to quake.

Earthquakes are very common events. Numbers on the map opposite mark the locations of thirteen earthquakes, all greater than magnitude 4.0 on the Richter scale, which happened on a single day: 5 March 2006. A magnitude 9.5 earthquake shook Chile in 1960, but by far the deadliest earthquake of modern times was the magnitude 9.0 earthquake which occurred on 26 December 2004 beneath the Indian Ocean, about 160 km (99 miles) west of northern Sumatra and about 30 km (18.6 miles) beneath the sea bed. That is when the ocean floor rose 13 m (43 feet) along over 500 km (310 miles) of the boundary between the Burma and Indo-Australian plates. The sudden shift sent waves through the ocean, travelling at over 800 km (500 miles) per hour. When they reached shallow water the waves slowed down but grew in height, piling the water into huge waves, known by their Japanese name as tsunamis. More than 220 000 people lost their lives. See pages 32–33 for more information on tsunamis.

Plates may move away from each other, toward each other, or they may pass each other in opposite directions. Plates are moving apart at ridges along the centres of each of the oceans. These boundaries are said to be constructive, because where two plates move apart hot rock rises from beneath the crust to fill the gap, cooling to form new crust. Where two plates collide, the boundary is destructive. If both plates are made from oceanic crust, the rocks from one plate slide beneath those from the other, but drag those rocks with them. The sinking rocks are said to be subducted into the Earth's mantle. When two continental plates collide both plates sink at the base, but the surface rocks are crumpled upward to form mountains. At a conservative margin, crust is neither constructed nor destroyed as the plates move past one another, jerkily, along a transform fault.

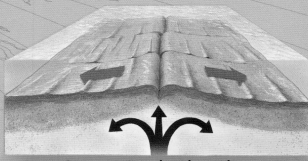

Constructive plate boundary

Destructive plate boundary (Oceanic)

Destructive plate boundary (Continental)

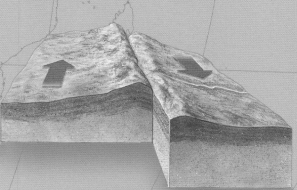

Conservative plate boundary

Earthquakes of magnitude 4 or more on 5 March 2006

The earthquakes listed in the following table are located on the map below.

Order	Time (GMT)	Location	Magnitude	Depth
1	2:40	Reykjanes Ridge, Atlantic Ocean	4.7	10 km (6.2 miles)
2	4:44	Jujuy, Argentina	4.7	191.4 km (118.9 miles)
3	5:06	Antofagasta, Chile	4.6	21 km (13 miles)
4	5:28	Northern Peru	5.3	114 km (70.8 miles)
5	8:07	Tonga	6.1	205.6 km (127.7 miles)
6	10:42	Northwest Territories , Canada	5.6	10 km (6.2 miles)
7	12:29	Rat Islands, Aleutian Islands, USA	4.0	15 km (9.3 miles)
8	12:43	Rota, Northern Mariana Islands	4.6	88.4 km (54.9 miles)
9	15:11	Tibet-Qinghai, China	4.6	11 km (6.8 miles)
10	17:12	Tonga	5.2	10 km (6.2 miles)
11	17:40	Sulawesi, Indonesia	4.5	178.1 km (110.6 miles)
12	20:46	Southern Sumatra, Indonesia	4.9	22.4 km (13.9 miles)
13	21:16	Southern Sumatra, Indonesia	5.0	29 km (18 miles)

The Earth's plates

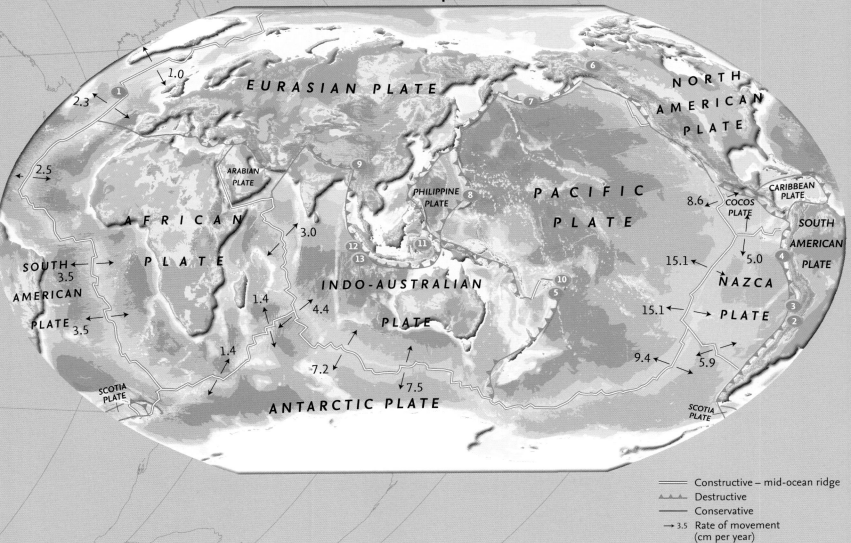

⎯⎯⎯	Constructive – mid-ocean ridge
▲▲▲	Destructive
⎯⎯⎯	Conservative
→ 3.5	Rate of movement (cm per year)
①	Earthquake location (see table above)

The citadel at Bam in Iran is 1000 km (630 miles) southeast of Tehran. It dates back 2000 years and is mainly made of mud bricks, clay, straw, and the trunks of palm trees. The city was originally founded during the Sassanian period AD 224–637 and its restoration had been

On 26 December 2003 an earthquake of magnitude 6.6 struck southeastern Iran killing over 43 000 people and destroying much of the city of Bam. About 60 per cent of the buildings were destroyed. The old quarter of the city and the citadel were severely damaged.

Ferry Building, San Francisco, USA 1906 and 2006

2006 is the centenary of the great San Francisco earthquake and fire. The top image shows the Ferry Building and its distinctive clocktower which survived the 1906 earthquake. It also survived the 1989 earthquake. The A. Page Brown steel-framed building originally opened in 1898 as a railway and ferry terminal. The lower picture shows it today. It is still a ferry terminal and also a thriving marketplace.

The earthquake and fire of 1906 destroyed buildings but not the overall street pattern of San Francisco. The main building in the top view looking down California Street is the Grace Church, which was subsequently relocated. The building on this site today is the Ritz Carlton Hotel. The relocated church became the Grace Cathedral, the third largest Episcopalian cathedral in the USA.

On 17 January 1995 Japan was struck by the Great Hanshin Earthquake of magnitude 7.2. Newer buildings had been built to withstand the seismic pressure and remained intact, but many residential buildings were traditional wooden structures which collapsed and burned. More than 5500 were killed as the epicentre was directly beneath a heavily populated region. If the quake had hit an hour later, during the morning rush hour, there could have been thousands more casualties. Today, buildings and roads are specifically designed to survive such events.

The spectacular collapse of the Hanshin expressway illustrates the effects of the high loads and strong ground movements the earthquake imposed on structures in the area. The city of Kobe has since been rebuilt, with a new expressway. However, many older residents were forced to move into the suburbs as typical Japanese insurance policies didn't cover earthquake damage. Many families lost everything.

Volcanoes

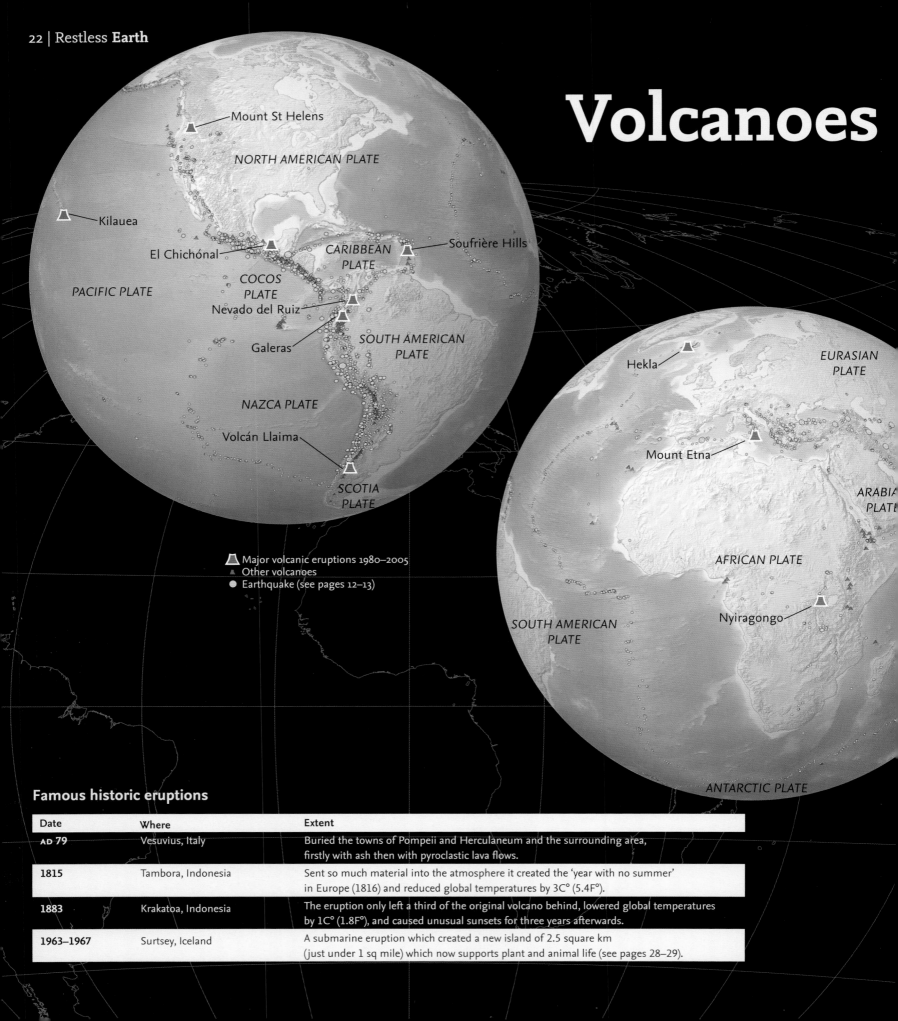

Mount St Helens

NORTH AMERICAN PLATE

Kilauea

El Chichónal

CARIBBEAN PLATE

Soufrière Hills

COCOS PLATE

PACIFIC PLATE

Nevado del Ruiz

SOUTH AMERICAN PLATE

Galeras

NAZCA PLATE

Volcán Llaima

SCOTIA PLATE

Hekla

EURASIAN PLATE

Mount Etna

ARABIAN PLATE

AFRICAN PLATE

SOUTH AMERICAN PLATE

Nyiragongo

ANTARCTIC PLATE

◭ Major volcanic eruptions 1980–2005
▲ Other volcanoes
● Earthquake (see pages 12–13)

Famous historic eruptions

Date	Where	Extent
AD 79	Vesuvius, Italy	Buried the towns of Pompeii and Herculaneum and the surrounding area, firstly with ash then with pyroclastic lava flows.
1815	Tambora, Indonesia	Sent so much material into the atmosphere it created the 'year with no summer' in Europe (1816) and reduced global temperatures by 3C° (5.4F°).
1883	Krakatoa, Indonesia	The eruption only left a third of the original volcano behind, lowered global temperatures by 1C° (1.8F°), and caused unusual sunsets for three years afterwards.
1963–1967	Surtsey, Iceland	A submarine eruption which created a new island of 2.5 square km (just under 1 sq mile) which now supports plant and animal life (see pages 28–29).

Major volcanic eruptions since 1980

Date	Volcano	Country
1980	Mount St Helens	USA
1982	El Chichónal	Mexico
1982	Gunung Galunggung	Indonesia
1983	Kilauea	Hawaii, USA
1983	Ō-yama	Japan
1985	Nevado del Ruiz	Colombia
1991	Mount Pinatubo	Philippines
1991	Unzen-dake	Japan
1993	Mayon	Philippines
1993	Galeras	Colombia
1994	Volcán Llaima	Chile
1994	Rabaul	Papua New Guinea
1997	Soufrière Hills	Montserrat
2000	Hekla	Iceland
2001	Mount Etna	Italy
2002	Nyiragongo	Democratic Republic of the Congo

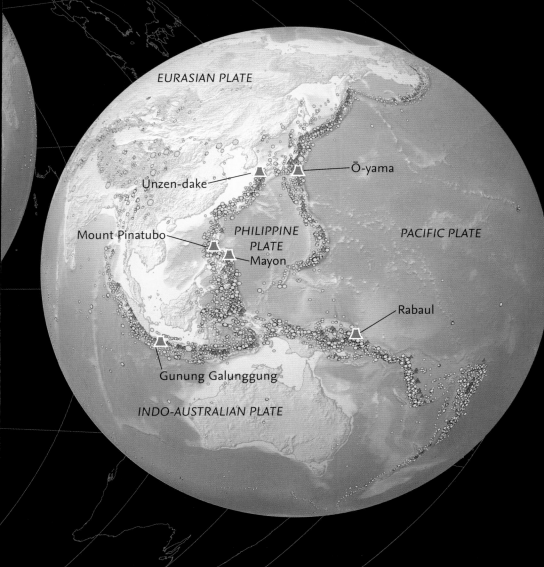

Beneath the solid rock of the Earth's crust lies the mantle, made from dense, hot rock which can flow very slowly. There are certain places where, from time to time, material from the mantle rises to the surface. As it rises the pressure compressing it relaxes and the material expands. Certain of its ingredients vaporize and may explode. Hot gas, molten rock, and blocks of solid rock rise to the surface and flow into the air or ocean. The hot material that is held under pressure just below the surface is called magma. When magma pours across the surface it is called lava. That is a volcanic eruption.

Volcanoes are found where the Earth's crust is thin and where there are fissures at the boundaries between the vast slabs, or plates, of rock which comprise the Earth's crust. They are particularly common around the shores of the Pacific Ocean, where active volcanoes form what is often called a 'ring of fire'.

many small explosions caused by the expansion of gases held inside very viscous lava. The lava is thrown upwards and falls back to build the cone. Not far from Stromboli there are volcanoes which produce Vulcanian eruptions. These explode with no release of magma, shattering the overlying solidified lava and hurling it high into the air, mixed with ash and gas. Mount St Helens, Mount Taranaki (Mount Egmont) in New Zealand, and Mount Fuji are strato-volcanoes, built up from layers of solidified lava alternating with layers of ash and loose rock. When they erupt they often do so explosively from a single vent. Shield volcanoes are broad cones with gently sloping sides. They are made from very liquid lava and their eruptions produce spectacular fire fountains. Many Hawaiian volcanoes are of this type.

Principal volcano types

Diagram	Type	Characteristics	Examples
	Fissure eruption	Lava erupts from linear cracks, very liquid and widespread flow.	Laki, Iceland
	Shield volcano	Very fluid basaltic lava erupts from a central vent, fast-flowing lava runs long distances from the vent.	Mauna Loa, Hawaii, USA Mauna Kea, Hawaii, USA
	Scoria or cinder cone	Explosive liquid lava emitted from a small central vent.	Paricutin, Mexico Stromboli, Italy
	Strato-volcano (Composite)	Viscous lava with explosive debris (pyroclast) erupts from large central vent.	Mount St Helens, USA Mount Taranaki (Mt Egmont), New Zealand Mount Fuji, Japan
	Lava dome with pyroclasts	Very viscous lava; relatively small, but can be explosive.	Mont Pelée, Martinique Soufrière Hills, Montserrat
	Caldera	Calderas form when large volumes of magma are lost, causing the formation of a large depression.	Kilauea, Hawaii, USA

Montserrat eruption chronology

21 September 1997 Major dome collapse to northeast of the volcano, destroying airport terminal building and entering the sea at various points along the coast.

3 August 1997 Major pyroclastic flows into Plymouth.

17–18 September 1996 First magmatic explosion with a major ash plume. 600 000 tonnes of ash was deposited in southern Montserrat.

3 April 1996 First pyroclastic flow, which travelled to the road which crosses the Tar River valley.

18 July 1995 Initial steam and ash venting from the northwest crater.

21 August 1995 First large eruption blankets Plymouth in a thick ash cloud and causes darkness.

12 May 1996 Pyroclastic flows reached the sea for the first time.

1–11 April 1997 Major pyroclastic flows down White River nearly reaching the sea at O'Garra's.

January 1992 Start of earthquake swarms in southern Montserrat.

25 June 1997 Major pyroclastic flows in Mosquito Ghaut reaching to within 50 m of the airport.

Peléean eruptions, named after Mount Pelée in Martinique which erupted violently in 1902 and more recently in 1995–1996, are often preceded by the accumulation of magma pushing the overlying rock upwards, forming a lava dome which then bursts. Such eruptions release huge clouds of extremely hot ash and small rocks in a pyroclastic flow which moves at great speed, burning and burying everything in its path. Volcanic ash consists of finely powdered rock which solidifies on contact with moisture. Inhaling it can be fatal. The eruption on Montserrat detailed on this page was also of this type.

If sea water floods into the top of an active vent it will cause a huge explosion, throwing a column of rock, ash, gases, and steam up to 20 km (12.4 miles) into the sky. This creates a Surtseyan eruption, a name coined in 1963 when an eruption of this type pushed a new island above the surface of the North Atlantic, to the south of Iceland. The island was named Surtsey and it is now alive with vegetation and birdsong (see pages 28–29. A Plinian eruption is even more violent, hurling ash and rock to a height of more than 50 km (31 miles). When Vesuvius erupted in this way in AD 79 it buried the city of Pompeii in ash, killing its inhabitants, including the writer Pliny the Elder, after whom this type of eruption is named.

Montserrat

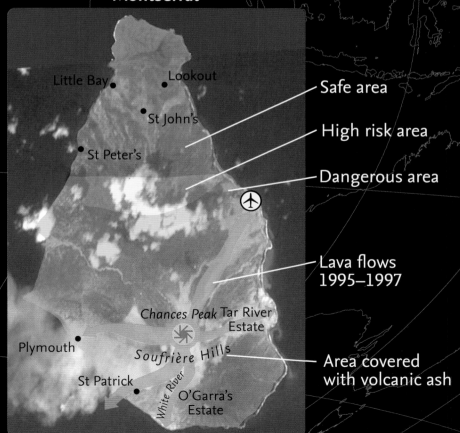

Little Bay
Lookout
St John's
St Peter's

Safe area

High risk area

Dangerous area

Lava flows 1995–1997

Chances Peak Tar River Estate

Plymouth

Soufrière Hills

St Patrick

White River

O'Garra's Estate

Area covered with volcanic ash

On 20 March 1980, after 123 years of inactivity, Mount St Helens awoke when an earthquake of magnitude 4.2 rumbled beneath it. Over the next two months a vast area was pushed outward which became known as 'the bulge'. This was caused by the rise of molten rock within the volcano. On 18 May another earthquake triggered a huge rockslide and a major volcanic eruption as the bulge gave way.

During the eruption, 400 m (1300 feet) of the north flank of the mountain collapsed or blew outwards and a plume of ash reached a height of 20–25 km (12–15 miles). The plume moved eastwards at an average speed of 95 km (60 miles) per hour, felling trees and killing most wildlife and vegetation within a 550 sq km (212 sq mile) area.

Birth of an island, Surtsey, Iceland November 1963

On 14 November 1963 fishermen observed the beginnings of an undersea eruption. An island was eventually created to a height of 169 m (554 feet) above sea level with an area of 2.5 sq km (0.97 sq miles). This island was named Surtsey after Surtur, the fire-possessing giant of Norse mythology who would set fire to the Earth at the Last Judgment.

Over a period of three and a half years Surtsey erupted and lava flowed. Before the eruption stopped the island was made a nature reserve and travel was restricted to scientists who studied the gradual development of life on this new land. Surtsey is now a favourite resting place for migratory birds and marine wildlife.

A town engulfed... Plymouth, Montserrat 1997

Plymouth, the former capital of the British territory of Montserrat in the Caribbean, was buried in volcanic ash as a result of eruptions of the Soufrière Hills volcano several times between 1995 and 1997. Its War Memorial was originally in a pleasant, grassy square but gradually became completely submerged in ash, as did its neighbouring telephone box.

Many buildings and facilities were buried and the town, which previously had a population of approximately 5000, was completely evacuated and abandoned as a result of the eruptions. The main damage was done by highly destructive pyroclastic flows carrying clouds of red hot material from the volcano. Only the far north of the island remained safe.

Tsunamis

Terrifying and destructive as earthquakes can be, sometimes they give rise to another phenomenon which can cause even more destruction and loss of life – the tsunami. When an earthquake occurs offshore, it may cause a sudden change in the shape of the ocean floor, as a result of vertical fault movement or submarine landslides. This causes a massive displacement of water, which in turn produces a powerful wave or series of waves capable of travelling over huge distances. Thankfully, although earthquakes are frequent occurrences, tsunamis resulting from them are relatively rare and it is even rarer for them to cause significant loss of life.

Tsunamis can travel at great speed in the open ocean. In the case of the Indian Ocean tsunami of 26 December 2004, the wave travelled at over 800 km (500 miles) per hour. In deep ocean water, the wave itself may seem small and insignificant, with heights of only 1 m (3 feet) greater than normal. However, as such waves reach shallower water their speed decreases but their height increases dramatically to create highly destructive waves which can be over 15 m (50 feet) high. The local coastal topography and shape of the sea bed influences the final effect of a tsunami, but the forces involved are enormous. It was the force of the water over vast areas – stretching far inland in some areas, and washing completely over smaller islands in others – which caused such widespread destruction and tragic loss of life in 2004.

Tsunamis in recent history

Date	Location	Extent	Caused by	Deaths
8 November 1929	Grand Banks of Newfoundland, Canada	Atlantic Ocean	Earthquake of magnitude 7.2 and submarine landslide	29
1 April 1946	Aleutian Islands, Alaska, USA	Pacific Ocean	Earthquake of magnitude 7.8	165
9 March 1957	Aleutian Islands, Alaska, USA	Pacific Ocean	Earthquake of magnitude 8.2	-
4 November 1957	Kamchatka Peninsula, Russian Federation	Pacific Ocean	Earthquake of magnitude 8.3	-
22 May 1960	South Central Chile	Pacific Ocean	Earthquake of magnitude 8.6	122
28 March 1964	Prince William Sound, Alaska, USA	Pacific Ocean	Earthquake of magnitude 8.6	-
29 November 1975	Hawaii, Pacific Ocean	Local	Earthquake of magnitude 7.2	2
26 December 2004	off Sumatra, Indonesia	Indian Ocean	Earthquake of magnitude 9.0	226 408

Cross-section of a tsunami

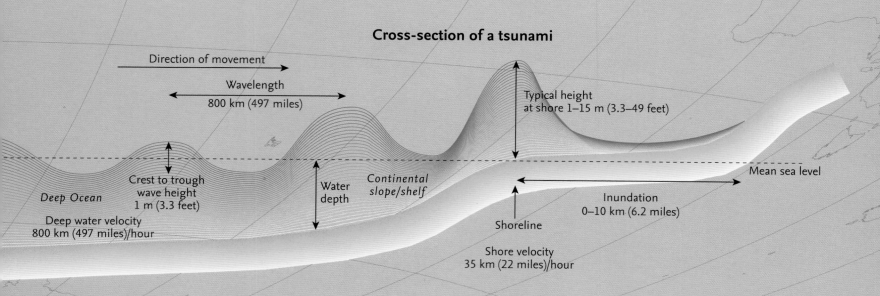

Direction of movement

Wavelength
800 km (497 miles)

Typical height
at shore 1–15 m (3.3–49 feet)

Crest to trough
wave height
1 m (3.3 feet)

Water
depth

*Continental
slope/shelf*

Mean sea level

Inundation
0–10 km (6.2 miles)

Deep Ocean

Deep water velocity
800 km (497 miles)/hour

Shoreline

Shore velocity
35 km (22 miles)/hour

Earthquake activity in Sumatra region

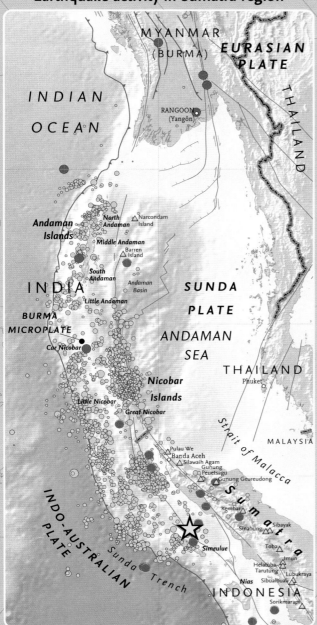

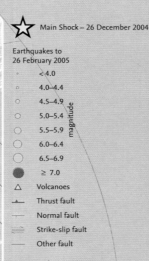

Main Shock – 26 December 2004

Earthquakes to
26 February 2005

magnitude

•	< 4.0
○	4.0–4.4
○	4.5–4.9
○	5.0–5.4
○	5.5–5.9
○	6.0–6.4
○	6.5–6.9
●	≥ 7.0
△	Volcanoes

	Thrust fault
	Normal fault
	Strike-slip fault
	Other fault

The most dangerous types of boundary between tectonic plates are those along subduction zones, where one plate is forced under another. Great pressure builds up over centuries as the plates converge. This pressure is released by the overlying plate slipping back into position when the rocks can no longer bear the pressure. In the Sumatran earthquake, it is estimated that the overriding Burma microplate shifted vertically by up to 13 m (43 feet), along a distance of over 500 km (310 miles). This vertical movement caused the sea to rise above the fault line, triggering the tsunami which then rapidly spread across the whole width of the Indian Ocean, reaching the coast of Somalia, over 5000 km (3100 miles) away, seven hours after the earthquake.

Indian Ocean tsunami death tolls

Estimated death tolls by country (March 2006)

Indonesia	165 708
Sri Lanka	35 399
India	16 389
Thailand	8 345
Somalia	298
Maldives	102
Malaysia	80
Myanmar	71
Tanzania	10
Seychelles	3
Bangladesh	2
Kenya	1
Total	226 408

Tsunami travel times (hours)

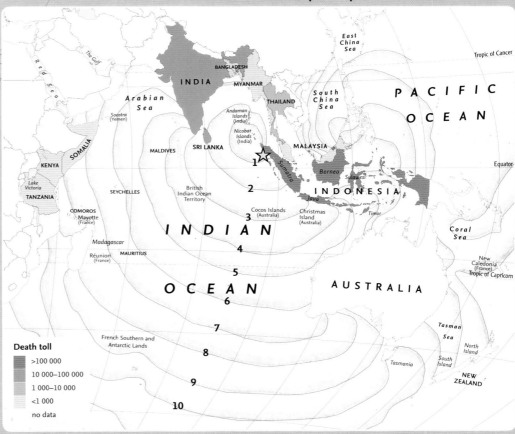

Death toll

	>100 000
	10 000–100 000
	1 000–10 000
	<1 000
	no data

Banda Aceh is a provincial capital of Indonesia on the very northern tip of the Indonesian island of Sumatra. The edge of the town is to the top right of this image. Seen here in a satellite image captured in January 2003, various types of agriculture are being undertaken and there is extensive woodland close to the town.

After the tsunami of 26 December 2004, much of the coast is under water as the surface vegetation and soil has been stripped off by the waves. In the town, buildings have been destroyed and a huge amount of debris has collected in some areas. Flooding extends well inland, where there is less destruction of property but widespread loss of agricultural land.

Lhoknga, Aceh, Indonesia 10 January 2003

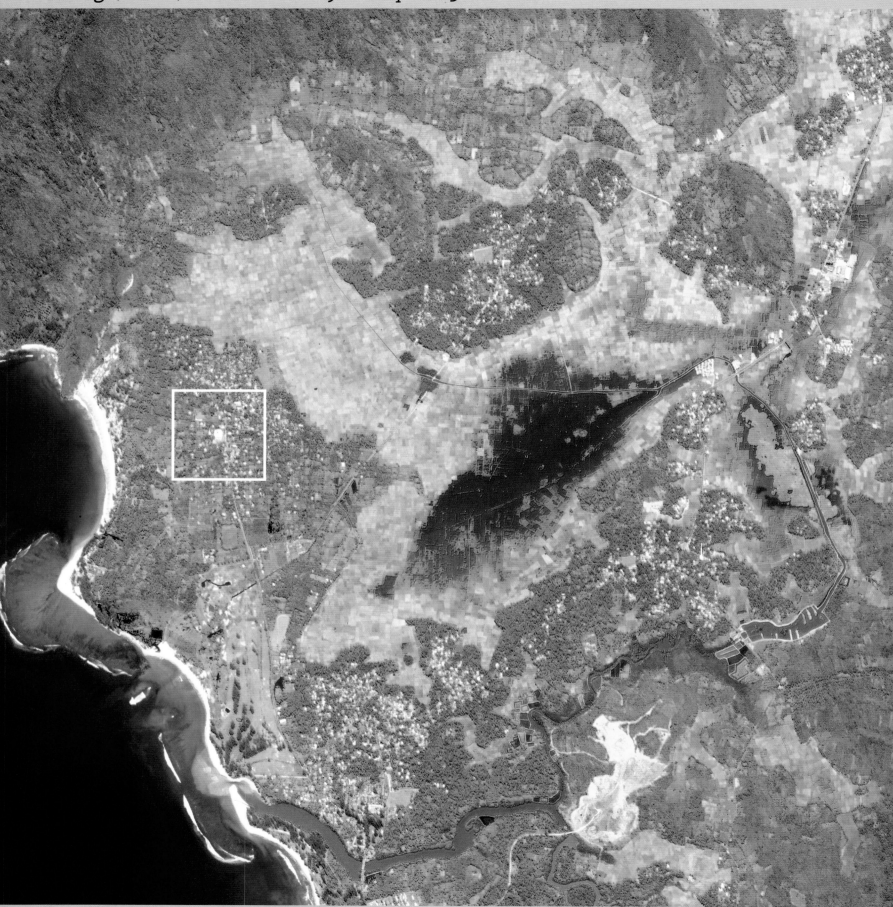

Captured in January 2003, this satellite image of Lhoknga near the provincial capital of Banda Aceh shows lush and well-cultivated land, with woodland and several villages. The darker area in the centre is water and there are several canals. The coast has sandy beaches, some with barrier islands or reefs protecting them.

Three days after the tsunami the extent of the destructive force of the waves can be seen. The coastal area has been stripped bare of vegetation and buildings with only the prominent Rahmatullah Lampuuk Mosque remaining (see page 39). Inland, the low-lying areas are now filled with salt water and it is only the slightly higher level of the roads which keeps them visible.

The 125-year-old Baiturrahman Mosque in Banda Aceh is an elegant building with 7 black domes and ornate architecture. Also known as the Grand Mosque, it survived the tsunami of 26 December 2004 but buildings closer to the coast were not so fortunate and thousands of people in the area lost their lives.

Immediately after the tsunami the Rahmatullah Lampuuk Mosque in Lhoknga, near Banda Aceh, was the only building left standing in the town. A vast area had been stripped of vegetation and housing (see pages 36–37). By 24 October 2005 vegetation had begun to regrow and some more permanent buildings had been built.

Deadliest landslides 1900–2006

Deadliest landslides 1900–2006

Date	Location	Country	Deaths
1934	Canton	China	500
1941	Huaray	Peru	5 000
1948	Assam	India	500
1949	Khait	Tajikistan	12 000
1955	Sumatra	Indonesia	405
1962	Ranrahirca area	Peru	2 000
1963	Vaiont Dam	Italy	2 600
1966	Rio de Janeiro	Brazil	350
1967	Rio de Janeiro	Brazil	436
1968	Bihar, Bengal	India	1 000
1971	Chungar	Peru	600
1972	West Virginia	USA	400
1973	Cholima	Honduras	2800
1973	Andes	Peru	500
1974	Mayunmarca	Peru	310
1987	Villatina, Medellin	Colombia	640
1995	Kulla, Himachal Pradesh	India	400
1995	Kurluk Zir Kutal, Badakhshan Province	Afghanistan	354
1997	Southern Peru	Peru	300
2002	Central/Eastern Nepal	Nepal	472
2006	Leyte	Philippines	139 (980 missing)

Steep slopes are often unstable. An earth tremor, heavy rain, or a storm may detach rocks or soil from the underlying solid rock. Gravity will do the rest, sending the loose material downhill as a landslide or, if rain has turned the soil to mud, as a mudslide.

Rock falls, landslides, and mudslides can bury entire villages, with huge loss of life. They can also cause disasters indirectly. In October 1963, ten days of heavy rain destabilized the side of Mount Toc, overlooking the reservoir behind the newly built Vaiont Dam in northern Italy. The surface of the mountain consisted of clay covered by loose rock and as the reservoir filled the rocks slid slowly downhill. Then the pressure of water from the saturated mountainside dislodged the rocks. Late on the evening of 9 October more than 240 million cubic metres (314 million cubic yards) of rock, sliding at more than 110 km (68 miles) per hour, fell into the reservoir. The splash sent huge waves about 300 m (984 feet) up the valley sides and water overflowed the dam wall. A wall of water 70 m (230 feet) high rushed down the valley, inundating four villages and killing more than 2500 people.

An avalanche is a mass of soil, rock, but more commonly snow, sliding down a slope. There is a risk of avalanches where a thick layer of snow accumulates on a slope of 30–40 degrees. There are two types of avalanche. A point-release avalanche affects only the surface layer and typically is shaped like an inverted V. A slab avalanche is more dangerous. It happens when an entire block of snow up to 800 m (2625 feet) wide is dislodged and moves at up to 200 km (124 miles) per hour, carrying everything before it. A slab avalanche also pushes air ahead of the snow, producing an avalanche wind blowing at up to 300 km (186.4 miles) per hour.

Landslides and avalanches

Anatomy of an avalanche

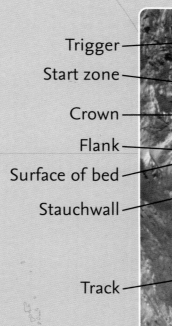

Trigger
Start zone
Crown
Flank
Surface of bed
Stauchwall
Track
Debris toe, Run-out

Landslide types

Type	Rate of movement	Material	Occurs along/in	Affected by
flow	rapid	loose soil, rock, water	steep slopes	surface water flow
fall	abrupt	rocks and boulders become detached	fractures, joints, bedding planes	gravity, weathering, water
topple	abrupt	forward rotation of units	pivotal point	gravity, force, water
slumps and slides	slow	soft rocks and sediments	planes of weakness	gravity, water,
creep	very slow and steady	soil or rock	areas of shear stress	water, temperature, material strength

Holbeck Hall Hotel, Scarborough, England 4 June 1993 and four days later

On 3 June 1993, Holbeck Hall Hotel was a four-star establishment standing about 65 m (213 feet) above the sea on the South Cliff, Scarborough on the East Yorkshire coast, UK. It looked out over an expanse of lawn to panoramic views of the North Sea. But by the 6 June, as a result of a massive landslide which took place in four stages, the lawn had disappeared and the ground had collapsed under the whole of the seaward wing of the hotel.

The rest of the hotel was unsafe and had to be demolished, but as the slip had been progressive everyone was evacuated without injury. Scarborough Borough Council was later held liable for the damage after being found to be in breach of its duty of care to maintain the supporting land.

Stieregg Hut, Grindelwald, Switzerland 1 June 2005 and two days later

The Stieregg Hut is a restaurant located at 1650 m (5413 feet) near Grindelwald in Switzerland. There was a massive landslide on Sunday 1 June 2005 which left the hut delicately positioned. Fortunately, the hut had been evacuated before the start of the summer season. Almost immediately there was another landslide and by 3 June the hut was precariously balanced, hanging over the edge of a very long and steep drop.

These satellite images show the Neelum river at Makhri just north of Muzaffarabad, before and after the magnitude 7.6 earthquake which struck northern Pakistan on 8 October 2005. Major landslides have blocked the river's usual course, forcing it to change direction. Its water is brown with sediment from many more landslides upriver.

Mount Cook, or Aoraki, is an imposing sight. In this image it is 3764 m (12 349 feet) high, but the steep slopes are unstable with hanging glaciers clinging to its slopes. On the night of 14 December 1991, climbing parties heard loud rumbles and saw bright sparks of rock impacts flashing in the dark.

Daylight revealed that the mountain had been dramatically changed. A great rock avalanche, estimated at 14 million cubic metres (18.3 million cubic yards) of the rock buttress, had travelled 7 km (4.3 miles) from the peak down the Hochstetter Icefall and Mount Cook was now 20 m (66 feet) shorter with a 500 m (1640 feet) scar. The debris from the avalanche stretched 3 km (1.2 miles) across the Tasman Glacier below the mountain.

Extreme

Storms

Wild weather

ropical storms – *severe and highly destructive storms created by intense low pressure weather systems, or cyclones, in tropical oceans*

ornadoes – *violent storms with very strong winds circling around small areas of extremely low pressure, characterized by tall, funnel-shaped clouds*

ust storms – *storms carrying fine particles into the atmosphere, common in areas of severe drought, causing poor visibility and loss of farmland*

now – *precipitation in the form of flakes of ice crystals formed in the upper atmosphere*

Tropical storms

Tropical cyclones, known as hurricanes in the Atlantic and Caribbean, typhoons in the Pacific and China Seas, and cyclones in the Bay of Bengal, begin as a disturbance in the distribution of air pressure. If the disturbance intensifies, generating sustained winds of up to 60 km (37 miles) per hour, it is classified as a tropical depression. When the winds increase beyond that it is called a tropical storm, and given a name. It becomes a tropical cyclone when it sustains winds of more than 120 km (75 miles) per hour. As the storm crosses the ocean, water evaporates into it, condensing to produce towering clouds. Condensation releases latent heat, warming the air and making it rise further. Evaporation and condensation supply the energy to drive the storm, and to generate a tropical cyclone the sea-surface temperature must be at least 24°C (75°F). That is why tropical cyclones develop only in the tropics and only during the summer. They also need the Coriolis effect, caused by the reaction of the atmosphere to the Earth's rotation, to make them turn. The Coriolis effect does not exist at the equator. Consequently, tropical cyclones cannot form closer than five degrees to the equator. Once formed, the storm travels westward until it approaches a continent. Then its track curves away from the equator. The Coriolis effect increases with distance from the equator, turning the track further.

2005 Hurricane tracks

NORTH AMERICA

Cindy

Ophelia

Nate

Rita

Katrina

Stan

Wilma

Beta

Dennis

Philippe

Maria

Emily

Epsilon

ATLANTIC OCEAN

Irene

SOUTH AMERICA

In late August 2005 hurricane Katrina swept across Florida, then crossed the Gulf of Mexico, intensifying as it did so. It made a second landfall near Buras, Louisiana, on 29 August and a third near the Louisiana-Mississippi border later the same day. As it crossed Lake Pontchartrain, the winds and torrential rain combined to breach, in three places, the levees protecting New Orleans. The resulting floods drove more than one million people from their homes and caused nearly 1400 deaths.

Hurricanes generate ferocious winds, but it is water, not wind, which causes the greatest damage and loss of life. As well as the heavy rain, the low pressure at the core of the storm allows the sea to bulge upward and the winds drive huge waves toward the shore. Together these produce a storm surge which can cause coastal flooding. Katrina produced a 9 m (30 feet) storm surge.

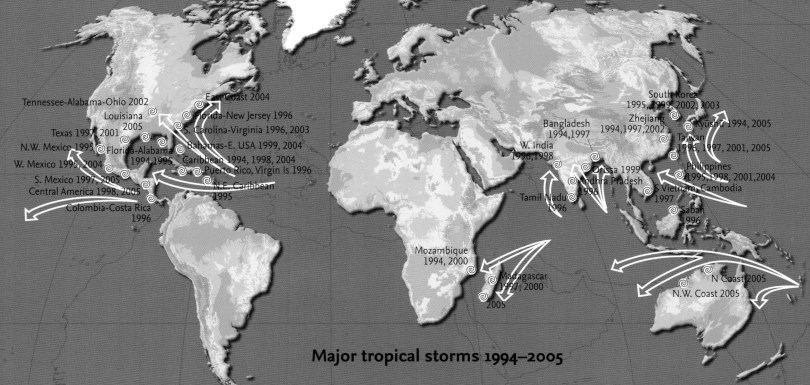

Major tropical storms 1994–2005

(Map labels)

Tennessee-Alabama-Ohio 2002
East Coast 2004
Louisiana 2005
Florida-New Jersey 1996
Texas 1997, 2001
S. Carolina-Virginia 1996, 2003
N.W. Mexico 1995
Florida-Alabama 1994, 1995
Bahamas-E. USA 1999, 2004
W. Mexico 1995, 2004
Caribbean 1994, 1998, 2004
S. Mexico 1997, 2005
Puerto Rico, Virgin Is 1996
Central America 1998, 2005
N.E. Caribbean 1995
Colombia-Costa Rica 1996

South Korea 1995, 1999, 2002, 2003
Bangladesh 1994, 1997
Zhejiang 1994, 1997, 2002
Kyūshū 1994, 2005
Taiwan 1996, 1997, 2001, 2005
W. India 1996, 1998
Orissa 1999
Andhra Pradesh 1996
Philippines 1995, 1998, 2001, 2004
Tamil Nadu 1996
S Vietnam, Cambodia 1997
Sabah 1996
Mozambique 1994, 2000
Madagascar 1997, 2000, 2005
N Coast 2005
N.W. Coast 2005

Atlantic and west Pacific tropical storms 1980–2005

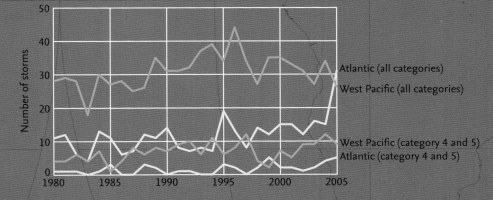

Number of storms (y-axis: 0, 10, 20, 30, 40, 50)
(x-axis: 1980, 1985, 1990, 1995, 2000, 2005)

Atlantic (all categories)
West Pacific (all categories)
West Pacific (category 4 and 5)
Atlantic (category 4 and 5)

The Saffir-Simpson hurricane scale

Tropical storms are reliant on warm water below them for energy and the slope of the continental shelf can affect the speed they hit land. Storms will abate when passing over land. The Saffir-Simpson scale ranks hurricanes from one to five in severity.

Category	Pressure		Wind speed			Storm surge above normal	
	kilopascals	millibars (mb)	km/h	mph	knots	m	ft
Tropical storm			63–118	39–73	34–63		
1	over 98	over 980	119–153	74–95	64–82	1.2–1.5	4–5
2	96.5–98	965–980	154–177	96–110	83–95	1.8–2.5	6–8
3	94.5–96.5	945–965	178–209	111–130	96–113	2.8–3.7	9–12
4	92–94.5	920–945	210–249	131–155	114–135	4–5.5	18
5	under 92	under 920	over 249	over 155	over 135	over 5.5	over 18

52 | Extreme **Storms**

Cross section of a tropical storm

Eye – cloudless and calm

Outflow of air at upper levels

Cumulo-nimbus cloud

Descending air in the eye

Rain bands

Increase in wind speed towards the centre

Moist air flows into the central area to replace rising air

Warm sea

Storm intensity is measured as the surface atmospheric pressure at the centre. While still over the ocean, Katrina's core pressure fell to 90.2 kilopascals (902 millibars). This made it the fourth most powerful storm ever recorded in the Atlantic and Caribbean. It was not the deadliest storm to strike the United States, however. That was the one which struck Galveston, Texas, in 1900, killing between 6000 and 12 000 people.

There were fifteen Atlantic hurricanes in 2005 and the official 1 June to 30 November hurricane season had to be extended into January 2006. This made 2005 a record-breaking year, but it came after a period of unusually low hurricane frequency and there have been worse. There were eighteen hurricanes in 1968 and twenty-one in 1933. Hurricanes are probably not becoming more frequent, and the winds they produce are not growing stronger, but the proportion of hurricanes in the fiercest category is increasing. This may be due to global warming.

The huge swirl of cloud of typhoon Longwan shows the typical form of a tropical storm. Originating in the Mariana Islands, it intensifed as it tracked westward and then northward towards the Philippines, developing into a typhoon. Within a day it had transformed its shape with clouds spiralling around the 'eye' in the centre. As it approached Taiwan it reached its peak strength before weakening as it finally made landfall in Fujian Province, China.

New Orleans, Louisiana, USA 28 August 2002 and 5 October 2005

Surrounded by levees built to keep the Mississippi from flooding the city, New Orleans was in a very vulnerable position. On a fateful day in October 2005 the unthinkable happened. Category 3 hurricane Katrina, having abated from category 5, struck the Louisiana coast making landfall at probaby the most vulnerable point – New Orleans. The combination of high winds and torrential rain resulted in the levees being breached in a number of places, resulting in widespread flooding. The satellite images above show the inundation alongside the Inner Harbor Navigation Canal (also shown in close-up on opposite page).

After the passage of Katrina the water began to drain back into the Inner Harbor Navigational Canal through the breach in the levee. The devastation of the flooding is evident in the top image. It resulted in the evacuation of thousands of residents as approximately 80 per cent of the city was inundated. The strength of the winds experienced during a severe tropical storm is shown in the lower image as category 5 hurricane Wilma whips up sand and bends palm trees as it roars into Miami Beach, Florida, USA.

A $13 million street renovation nears completion in May 2005 with improvements to the sidewalks and street lighting, and the planting of over 200 Moroccan palms, which cost $7000 each. When hurricane Katrina swept through the city this important tourist area within the city

The effects of the storm surge which can accompany a tropical storm as it makes landfall can be seen in this pair of images of Gulfport, Mississippi, USA. Most noticeable are the three-storey barge which has been left high and dry in the container terminal, and the containers washed inland by the surge, estimated to have been 8–9 m (28–30 feet), a record for an Atlantic storm. It was estimated that the damage to

New Orleans, Louisiana, USA 30 August 2005

As a result of Katrina 80 per cent of New Orleans was under water and the storm was estimated to have been the costliest and one of the deadliest hurricanes in the history of the USA. Much of the city was evacuated, many areas were left uninhabitable and residents were accommodated in neighbouring cities and towns in temporary homes.

As the flood waters receded there was almost a scene of normality except for the lack of vehicles on the roads. However, cars remain abandoned on the roads as residents were not permitted to return to the city immediately after the storm. Six months after the storm, nearly o per cent of the 272 km (169 miles) of breached levees and floodwalls had been repaired. The target was to complete this mammoth task before the start of the 2006 hurricane season.

Tornadoes

Fujita tornado scale

F-scale number	Wind speed		Type of damage done
	km/h	mph	
F0	64–116	40–72	Light damage. Some damage to chimneys; branches broken off trees; shallow-rooted trees pushed over; sign boards damaged.
F1	117–180	73–112	Moderate damage. Peels surface off roofs; mobile homes pushed off foundations or overturned; moving cars blown off roads.
F2	181–253	113–157	Considerable damage. Roofs torn off frame houses; mobile homes demolished; large trees snapped or uprooted; light-object missiles generated; cars lifted off ground.
F3	254–332	158–206	Severe damage. Roofs and some walls torn off well-constructed houses; trains overturned; most trees in forest uprooted; heavy cars lifted off the ground and thrown.
F4	333–419	207–260	Devastating damage. Well-constructed houses levelled; structures with weak foundations blown away some distance; cars thrown and large missiles generated.
F5	420–512	261–318	Incredible damage. Strong frame houses levelled off foundations and swept away; automobile-sized missiles fly through the air to heights in excess of 100 m (328 ft); trees debarked; incredible phenomena occur.
F6	513–610	319–379	Winds of these speeds are unlikely

Tornadoes in the USA

40 — Average number of tornadoes per year 1950–1998
40 — Average number of tornadoes per year 1989–1998
Areas with high frequency of tornadoes
Major tornadoes

One wet and windy Saturday afternoon in October 2000 a tornado carved a swathe of destruction more than 2 km (1.2 mile) long through the English seaside resort of Bognor Regis. A man and woman driving along the road suddenly encountered a wall of flying debris. Then their car rose into the air, drifted to the opposite side of the road, and landed with a bump. Tornadoes can and do happen anywhere, although they are least common in the tropics. The United States suffers from tornadoes more than any other country. There were approximately 150 in the first three months of 2006, and on 2 April 2006, an outbreak of sixty-four tornadoes affected an area from Iowa to Mississippi and from western Missouri to the Ohio Valley. Tornadoes are most frequent in an area known as Tornado Alley, centred on Texas, Oklahoma, and Nebraska, but encompassing parts of seven other states.

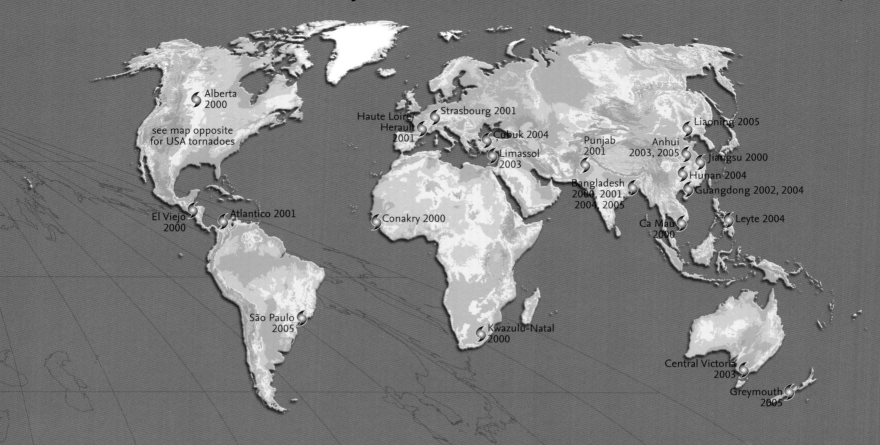

Alberta 2000

see map opposite for USA tornadoes

Haute Loire/ Herault 2001

Strasbourg 2001

Çubuk 2004

Limassol 2003

Liaoning 2005

Punjab 2001

Anhui 2003, 2005

Jiangsu 2000

Hunan 2004

Guangdong 2002, 2004

Bangladesh 2000, 2001, 2004, 2005

Ca Mau 2000

Leyte 2004

El Viejo 2000

Atlantico 2001

Conakry 2000

São Paulo 2005

Kwazulu-Natal 2000

Central Victoria 2003

Greymouth 2005

USA tornadoes and associated deaths 1950–1999

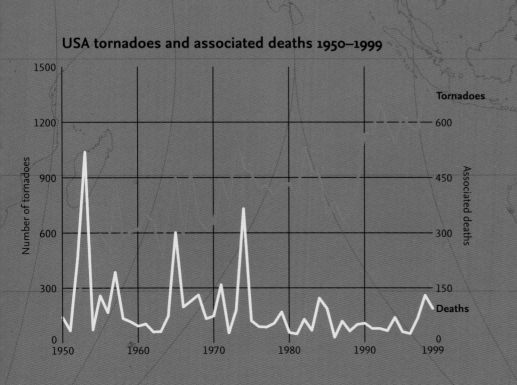

A tornado is a rapidly rotating column of air spiralling upward. The biggest and fiercest of these 'twisters' develop inside huge storm clouds containing a single convection cell, called a supercell. If air inside the cloud starts to rotate, the rotation can extend downward through the cloud, emerging from the base as a funnel, visible because the low pressure causes water vapour in the air to condense. The funnel becomes a tornado when it touches the ground and a waterspout if it touches water. Less violent conditions sometimes produce weak, short-lived tornadoes. Air accelerates as it is drawn into the tornado vortex. Most tornadoes generate winds of less than 200 km (124 miles) per hour – sufficient to cause serious damage – but occasionally the wind speed may reach 500 km (311 miles) per hour. The number of tornadoes appears to have increased dramatically over the last half century. This is due entirely to improved detection and reporting, however. There is no evidence that tornadoes are becoming more frequent.

This dramatic series of images shows the formation of a waterspout in the Adriatic Sea off the coast of Croatia. The wind speed on the edge of this waterspout reached 170 km (66 miles) per hour but it caused little damage, however, as it did not reach land. In the first image the air within the cloud has started to rotate and is emerging from it as a funnel. Gradually this funnel extends downwards until it reaches the

surface of the sea and the waterspout is formed. If this had taken place over land it would have become a tornado with a likelihood of damage occurring. In the last two images a second funnel is beginning to form. It is common for one weather system to spawn a number of tornadoes, resulting in damage over a considerable area as the system tracks across country.

Tornado damage, USA August 1999 and April 2006

The destructive power of a tornado is illustated well by the above images. The upper one shows debris being swirled through the air by a tornado which hit Salt Lake City, USA. The lower shows the track and the resulting severe damage in Gallatin, Tennessee caused by a tornado in April 2006. The narrowness of the track is emphasised by the relatively untouched buildings on both sides.

Tornadoes take a variety of forms and names – twister, whirlwind, wedge, funnel – and can vary considerably in size. No two are the same and they are difficult to predict. The ghostly appearance of this tornado is caused by the lighting conditions at the time and by debris being thrown up into the air.

The trail of damage left by one of a number of large tornadoes which hit the Great Plains region of the United States on 3 May 2003 can be clearly seen in this satellite image of Moore, Oklahoma taken a week after the event. Across the region more than seventy tornadoes were recorded, resulting in forty fatalities. Within the counties of Oklahoma and Cleveland nearly 1800 homes were destroyed and over 6500 damaged.

The tornado which struck Henderson, Kentucky completely wrecked houses and mobile homes and resulted in over twenty deaths. The swirl marks in the field adjacent to the remains of these farm buildings illustrate the spiralling nature of the storm and how it scoured the earth as it passed through. Note also how the building at the bottom right of the picture is relatively undamaged, yet only a very short distance away another building has been completely destroyed.

Dust and sand storms in China

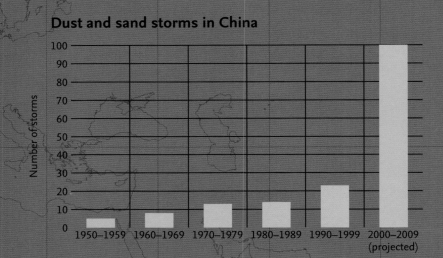

Even a light wind is enough to blow dry dust into the air and a wind of 20 km (12 miles) per hour will lift dry sand grains. Usually the dust and sand will soon settle, but if the wind blows the particles into air that is rising, they may be carried to a great height. And if the supply of dust or sand is virtually limitless, the combination of a strong wind and rising air can produce a dust storm or sand storm – the difference being only in the size of the particles. A severe storm advances as a swirling wall of dust which reduces visibility almost to zero. Dust and sand penetrate clothes and enter buildings despite the doors and windows being tight shut.

In May 1934, during the American drought which produced the Dust Bowl, dust settled on President F. D. Roosevelt's desk in Washington as fast as it could be wiped away. Ducks and geese choked and fell dead from the sky. At one time a cloud of dust – prairie soil – 5 km (3 miles) high extended from Canada to Texas and from Montana to Ohio. Once airborne, dust can travel long distances. Saharan dust quite often reaches Britain, 2500 km (1554 miles) away, and sometimes it crosses the Atlantic where rain washes it down onto American cars. Beijing suffers every spring when dust blows over the city from the Gobi Desert. Nearly one million tonnes of Gobi dust fall on Beijing every year. The expansion of agriculture into marginal lands and the lowering of the water table due to extraction of water for irrigation and industrial use have left the soil exposed and dry. Consequently, the Chinese dust storms are becoming more frequent and more severe. The dust spreads to neighbouring countries and as far as Hawaii. In response, the Chinese have planted millions of trees to protect the capital.

Dust storms

Frequent dust storm tracks

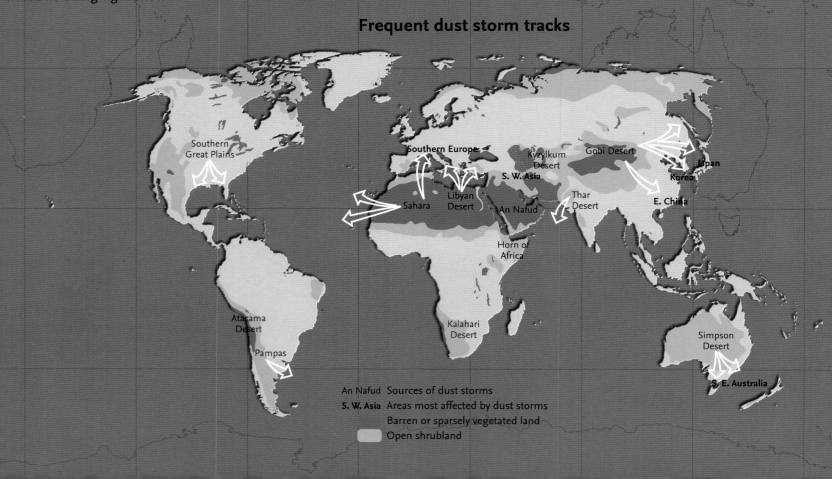

An Nafud — Sources of dust storms
S. W. Asia — Areas most affected by dust storms
Barren or sparsely vegetated land
Open shrubland

A dust storm sweeps in an anticlockwise direction eastward and northward from Libya, over Egypt and into the Mediterranean. The island of Crete is almost invisible below the sand and dust which originated in the Sahara desert far to the south. The low pressure system towards the top left of the image is carrying the dust into mainland Greece and Turkey. Saharan dust is sometimes transported as far north as the United Kingdom.

Al Asad, Iraq   26 April 2005

A major hazard encountered during the recent war in Iraq has been the common occurrence of dust storms. This one originated near the Syria and Jordan border and moved eastward, increasing in strength and creating a wall of dust reducing visibility to almost zero. It was estimated that the wall of dust may have reached a height of 1220–1524 m (4000–5000 feet).

When the Spring winds blow from the Gobi Desert they ofter carry large amounts of sand and dust seastward towards Beijing and onwards to the Korean peninsula and even as far as Japan. In Beijing the sky turns yellow with the dust in the atmosphere, visibility is greatly reduced and the inhabitants have to wear masks to aid with breathing.

Dust storms are by no means a new phenomenon. In the 1930s the Great Plains region of North America was notorious for its frequent dust storms and became known as the 'Dust Bowl'. As a result of poor farming techniques and drought the soil turned to dust and was carried away eastward by the wind, reaching as far as the Atlantic Ocean. In 1935 these two photographs were taken in Kansas, one of the many

states to suffer during the Dust Bowl era. Taken only fifteen minutes apart, it is only the street lights which confirm that the images are of the same scene.

Snow

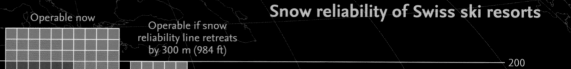

In January 1996, winter storms which had brought blizzards to the eastern United States crossed the ocean and struck Britain. All the roads to the Sellafield nuclear reprocessing plant in Cumbria were blocked with snow and workers had to spend two nights in the plant because they could not get home. Hundreds of motorists were trapped in snow on the roads of southwestern Scotland. Snowstorms are highly disruptive. In 2003 snowstorms paralysed large parts of the northeastern United States. Airports closed, rail services were brought almost to a standstill, and roads were not only blocked, but in some places the snow hid them completely. For a short time in March 2006, snow blocked all the roads between England and Scotland.

If snowstorms are bad, blizzards are worse. Technically, a blizzard is defined as a wind of at least 56 km (35 miles) per hour with snow falling heavily enough to deposit a layer at least 25 cm (10 inches) thick or with visibility reduced to less than 400 metres (1312 feet). People rapidly become disorientated in such whiteout conditions and may die of cold while seeking shelter. Mountain resorts welcome the snow, of course, but the quality of snow depends on the temperature inside the clouds which produce it. The best snow for skiing is dry and powdery, and falls from the coldest clouds. As global temperatures rise, however, there may be fewer winters with good snow for winter sports.

Snow reliability of Swiss ski resorts

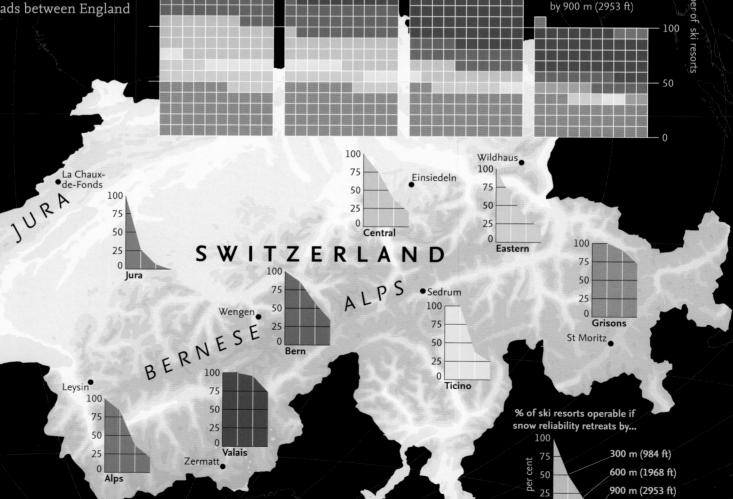

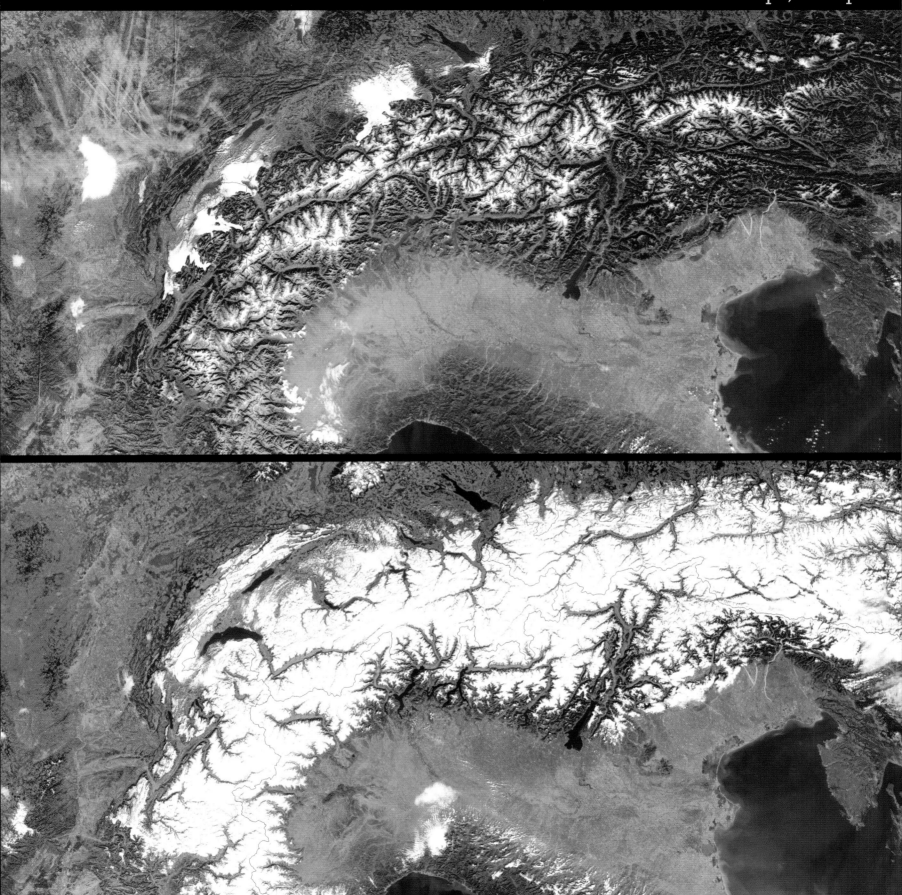

Winter in the Alps, Europe

In November, at the beginning of winter, only the highest peaks and ridges are capped in snow with some having remained snow-covered all year. However, towards the end of the winter the increase in snow cover can be clearly seen on the lower satellite image taken in March. Most of the Alpine region is now covered in snow with only the lower valleys being clear.

Winter in the northern hemisphere

Snow cover in January in the northern hemisphere extends in an almost horizontal band as far south as the USA/Canada border and from eastern Europe to northern China. South of this line the high mountain ranges of the Rocky Mountains, the Caucasus, the Himalaya and

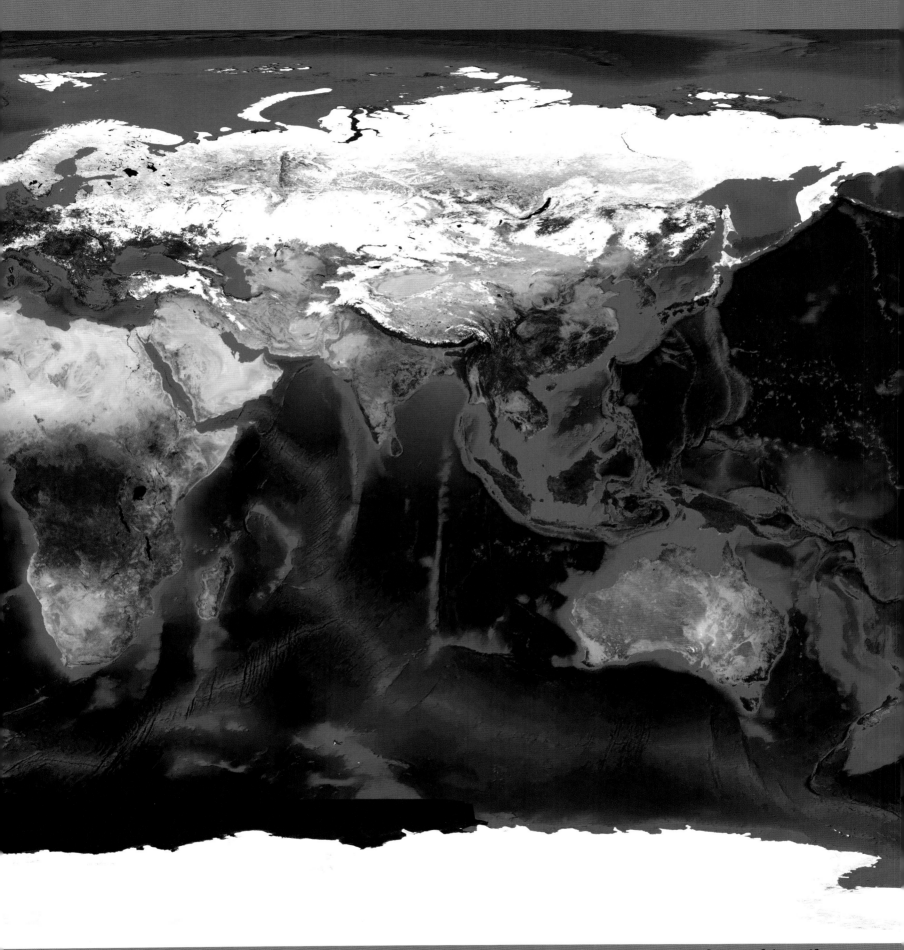

Japan are snowclad. Western Europe, as a consequence of its position next to the Atlantic Ocean and the warming influence of the Gulf Stream, is snow-free. The southern hemisphere, apart from the Antarctic continent, is virtually clear of snow.

Winter in the southern hemisphere

In July the difference in snow cover in the northern hemisphere is immediately obvious with snow now confined to the Arctic regions and the highest mountain ranges. Even though it is now winter in the southern hemisphere there is little evidence of snow except in the Andes in South

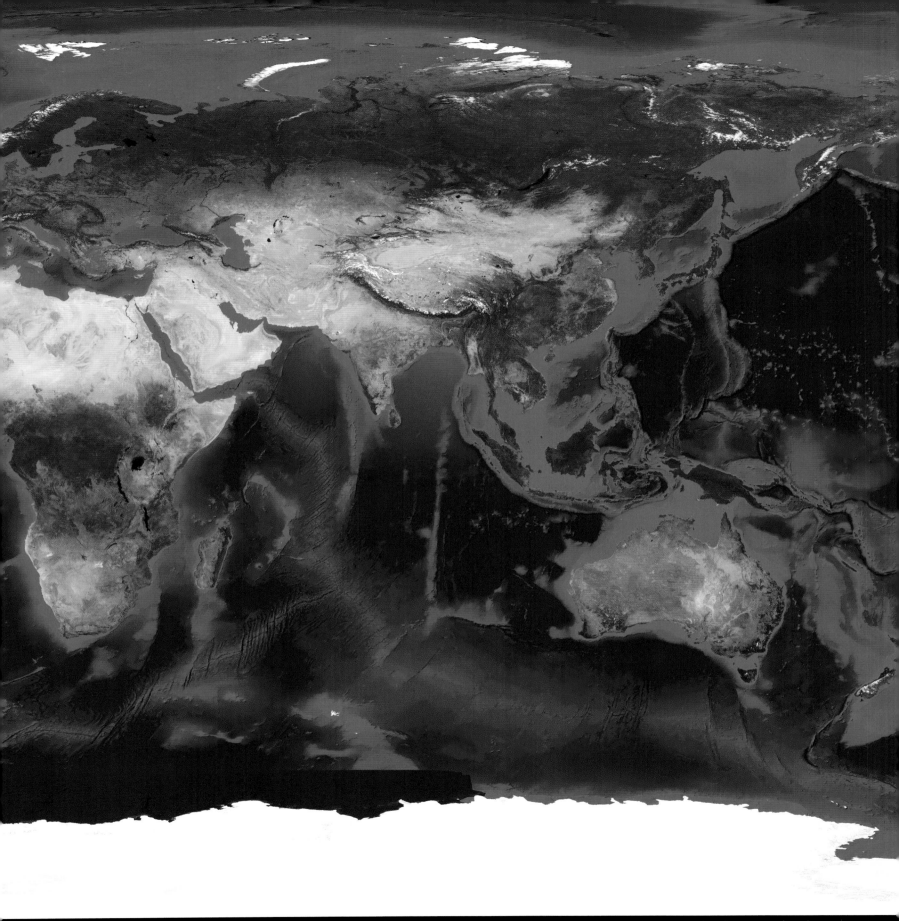

America and the Southern Alps of New Zealand. The land mass of Antarctica continues to be covered but as there is no land immmediately to its

Man-made

World

Building and development

creating land – *the process of reclaiming land from the sea, or converting the use of derelict or flooded areas, to create productive land*

controlling water – *use of water resources for the generation of power and for the irrigation of otherwise infertile land for agriculture*

expanding cities – *outward and upward growth of urban areas largely in response to the migration of people from rural areas*

travel and transportation – *creation of roads, railways and transport networks for the movement of goods and people throughout regions and around the world*

Creating land

Two-thirds of the planet is covered by sea. Every year sea-level rise takes a little more. But in some places mankind is fighting back by draining lagoons and filling in shallow coastal waters with dredgings and waste materials. The first modern masters of land reclamation were the Dutch. A quarter of the Netherlands' land area is below sea level, mostly made up of 3000 reclaimed 'polders'. The largest, including Flevoland, the world's largest man-made island, are within the IJsselmer, a large lake created in the 1930s by barricading off a large bay, the Zuider Zee, from the North Sea.

Cities built on drained coastal lagoons, swamps and harbours include Venice, St Petersburg and parts of Boston and Tōkyō. Most recent large reclamation projects have been in Asia. Hong Kong has created new flat land in its harbour which now houses much of the central business district. It also recently extended Lantau Island to accommodate its new international airport. Nearby Macao airport is also built on former sea bed, as is Japan's second-largest international airport at Kansai near Osaka.

Netherlands Zuider Zee project

Waddenzee

North Sea

Wieringenmeer
1927–1930

IJsselmeer

Northeast
Polder
1937–1942

Markermeer

Lelystad ●

Eastern
Flevoland
1950–1957

Southern
Flevoland
Almere ● 1959–1968

AMSTERDAM

● New town
— Dykes
— Old coastline, 1920s

IJsselmeer polders	Area sq km	sq miles	% of total land area of Netherlands	Period of construction	Agriculture	% land use Woodland	Residential	Other
Wieringermeer	200	48	0.6	1927–1930	87	3	1	9
Northeast polder	480	115	1.4	1937–1942	87	5	1	7
Eastern Flevoland	540	130	1.6	1950–1957	75	11	8	6
Southern Flevoland	430	103	1.3	1959–1968	50	25	18	7
Total	1 650	396	5.0	1927–1968	73	12	8	7

China is constructing the world's largest container port on reclaimed land round an island off Shanghai. The island state of Singapore has reclaimed land around sixty offshore islands, creating over 100 sq km (39 sq miles) of new land, though at the loss of several valuable coral reefs. Most dramatically, the fast-growing city of Dubai is currently building entire archipelagos of small artificial islands in The Gulf. Made from material dredged from the Gulf shipping lanes, these archipelagos will house luxury hotels, beach-front homes and holiday villas, and marinas.

But not all land reclamation goes smoothly. Kansai airport in Ōsaka, Japan is sinking. Malaysia has appealed to the United Nations over Singapore's 'reclaiming' of parts of the narrow Johor Strait between them. And the Netherlands is contemplating giving some of its polders back to the sea, because of the cost of keeping them drained.

Singapore - growth

Date	Area sq km	sq miles
1963	580	224
1988	625	241
2003	697	269
Future target	760	293

Singapore island land reclamation

The coastline as it appeared in 1958 is superimposed in yellow. ——

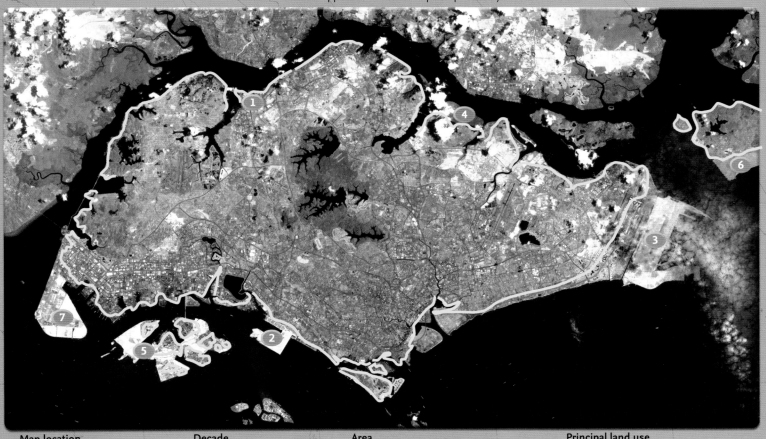

Map location	Decade	Area	Principal land use
1	1960	Kranji	industrial
2	1970	Pasir Panjang	port
3	1970	Changi	international airport
4	1980	Seletar	housing
5	1990	Jurong Island	petrochemicals
6	2000	Pulau Tekong	*proposed new towns*
7	2000	Tuas	*proposed port*

In the 1930s the Dutch began the Zuider Zee works to reclaim a large area of land from the sea. In the above satellite images taken in 1964 and 1973 Flevoland is under reclamation. In 1973 a dyke is under construction to create the southernmost polder, part of which was to be taken up by a new town - Almere. The northern part of Flevoland is fully drained and under cultivation and includes the new town of Lelystad.

By 2004 the reclamation of Flevoland is complete and the polders are all under cultivation. The area between Almere and Lelystad was not fully drained and has become an important nature reserve for waterfowl. The lighter blue water is Markermeer, an undrained polder which serves as a freshwater reservoir and flood control.

Palm Island, Dubai February 2002, November 2002, June 2003, September 2004

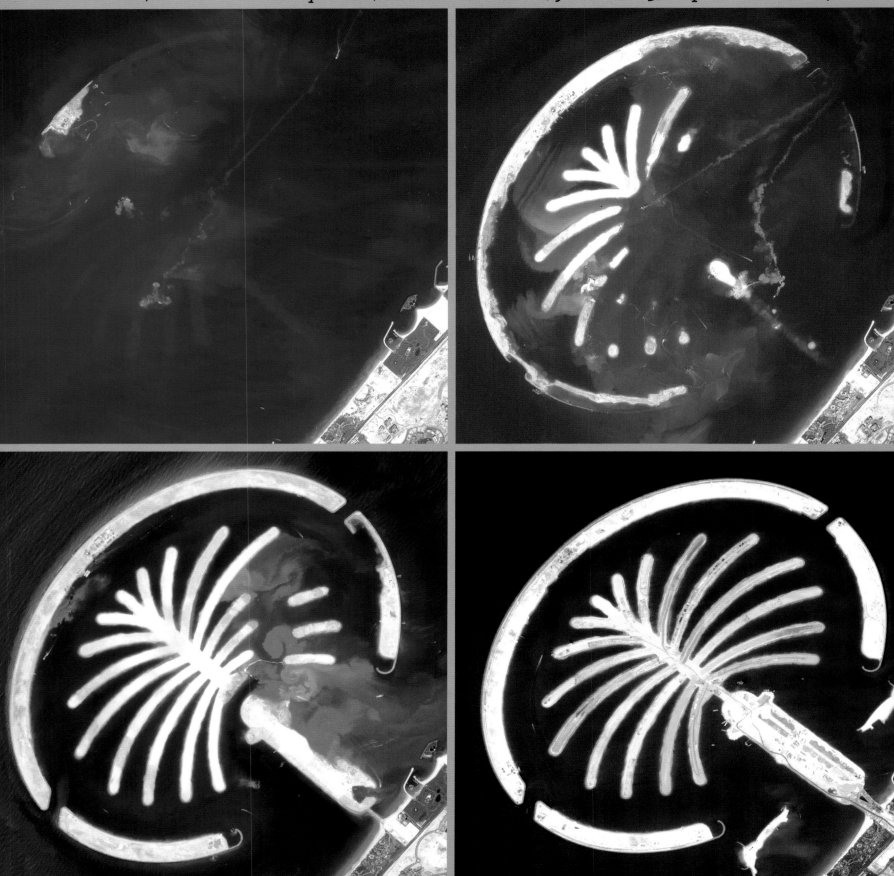

Work commenced at the beginning of the millennium on the world's largest land reclamation project to create a series of man-made islands which would become luxury resorts on the coast of Dubai in the United Arab Emirates. This sequence of images illustrates how the first 'Palm Island' gradually took shape as reclamation progressed. When complete, the 'island' will contain 2000 villas, 40 luxury hotels, shopping complexes and other facilities.

But that is only the first part. This image taken in 2005, shows a second island almost complete and the start of the third and largest of the palms at the top of the picture. However, even these developments seem small when compared to 'The World', a series of 300 islands which together will give the appearance of the continents. Already North America, Asia and Africa are taking shape. Transportation between individual islands will be by sea and air.

Hong Kong has been reclaiming land from the sea for over 50 years to accommodate an ever-increasing population and expanding commercial activity. One of the most recent developments has been in West Kowloon where the land has been reclaimed for a variety of uses including commercial, residential and recreational. In the lower image the road link to the new international airport, built on reclaimed land off neighbouring Lantau Island, can be seen entering the tunnel under Victoria Harbour.

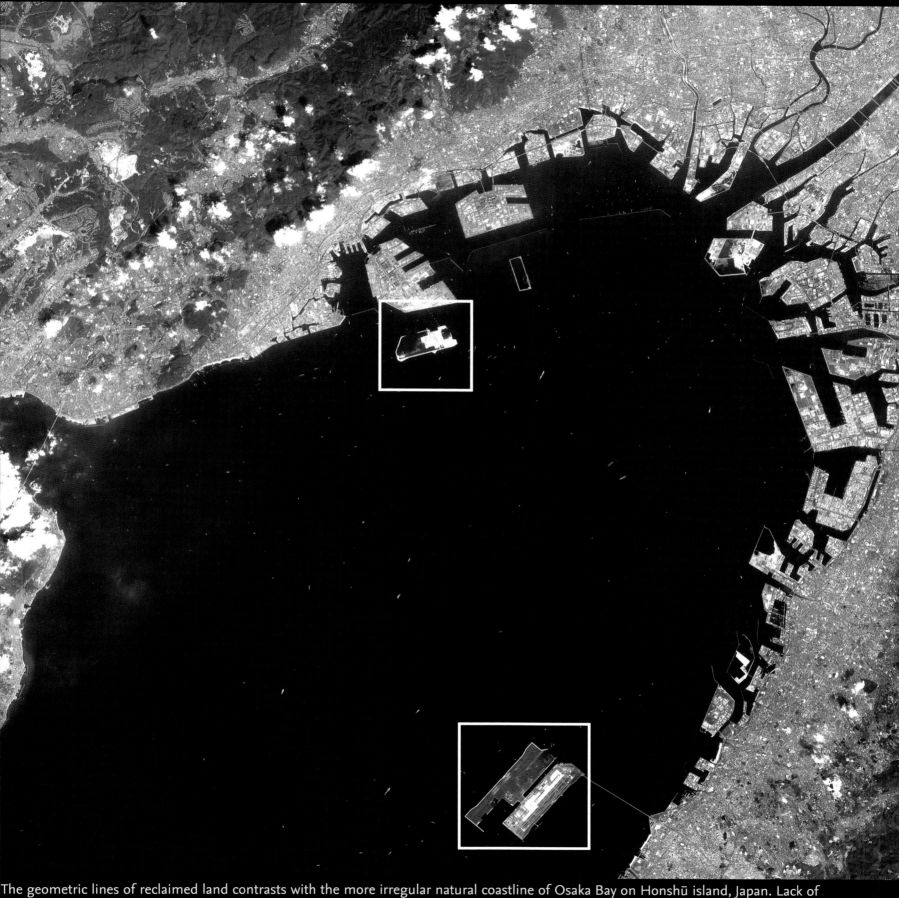

The geometric lines of reclaimed land contrasts with the more irregular natural coastline of Osaka Bay on Honshū island, Japan. Lack of suitable land for urban expansion has forced the Japanese to reclaim a large amount of land from the sea. Around Osaka Bay this new land is used mostly for commercial and port enterprises with the two new Kansai International and Kobe airports indicated in the image.

Only a handful of truly wild rivers remain in the world – mostly in the Arctic and in a few remote rainforests. Virtually all the rest are tamed by dams and dykes which aim to prevent floods, irrigate crops, fill taps or generate electricity.

Since the construction of the 220 m-(722 feet) high Hoover Dam on the River Colorado in the USA in the 1930s, engineers have built hundreds of giant dams worldwide. These generate one-fifth of the world's electricity. China's new Three Gorges Dam on the Yangtze is the largest yet, and will generate as much power as twenty coal or nuclear power stations.

Controlling water

Highest dams

	Country	Height m	feet
Rogun	Tajikistan	335	1 100
Nurek	Tajikistan	300	984
Xiaowan (Yunnan Gorge)	China	292	958
Grand Dixance	Switzerland	285	935
Inguri	Georgia	272	892
Vaiont	Italy	262	860
Manuel M. Torres	Mexico	261	856
Tehri	India	261	856
Álvaro Obregón	Mexico	260	853
Mauvoisin	Switzerland	250	820

Largest volume embankment dams

	Country	Volume (thousands) cubic m	cubic feet
Tarbela	Pakistan	148 500	194 238
Fort Peck	USA	96 050	125 633
Tucuruí	Brazil	85 200	111 442
Atatürk	Turkey	85 000	111 180
Yacyretá	Argentina	81 000	105 948
Rogun	Tajikistan	75 500	98 754
Oahe	USA	70 339	92 003
Guri (Raúl Leoni)	Venezuela	70 000	91 560
Parambikulam	India	69 165	90 468
High Island West	China (Hong Kong)	67 000	87 636

Major hydro-electricity plants and irrigated land

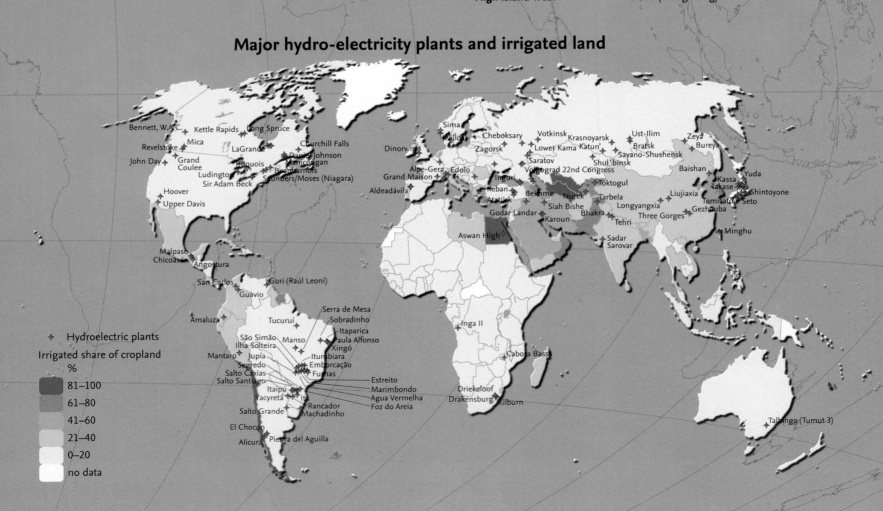

Hydroelectric plants

Irrigated share of cropland
%

81–100
61–80
41–60
21–40
0–20
no data

Largest capacity reservoirs

	Country	Capacity (millions) cubic m	cubic feet
Kariba Gorges	Zimbabwe/Zambia	180 600	236 225
Bratsk	Russia	169 000	221 052
Aswan High	Egypt	162 000	211 896
Akosombo	Ghana	147 960	193 531
Daniel-Johnson	Canada	141 851	185 541
Guri (Raúl Leoni)	Venezuela	135 000	176 580
Bennett, W.A.C.	Canada	74 300	97 184
Krasnoyarsk	Russian Federation	73 000	95 484
Zeya	Russian Federation	68 400	89 467
La Grande 2	Canada	61 715	80 723

Largest hydroelectric plants

	Country	Planned generating capacity (MW)
Three Gorges Dam (Sanxia)	China	18 200
Itaipu	Brazil	12 600
Guri (Raúl Leoni)	Venezuela	10 000
Tucuruí	Brazil	8 370
Sayano-Shushensk	Russia	6 400
Itaipu	Paraguay	6 300
Krasnoyarsk	Russia	6 000
Bratsk	Russia	4 500
Longtan (Guanxi, Tian'e)	China	4 200
Xiaowan (Yunnan)	China	4 200

Meanwhile, large dams have allowed governments to triple the amount of water delivered to farmers to irrigate crops in the past forty years, enabling the world to feed its fast-growing population. More than two-thirds of all the water used by humans supplies irrigated fields which grow almost 40 per cent of the world's food – and more than 80 per cent in countries such as Egypt and Pakistan.

But there is a big downside. In 2000, experts on the World Commission on Dams reported that many dams have caused environmental and social havoc which outweighs the economic gains. Dams have flooded fertile valleys and forests, created millions of refugees, destroyed fisheries, emptied wetlands, eroded river banks and caused flooding.

Three Gorges Dam project, China

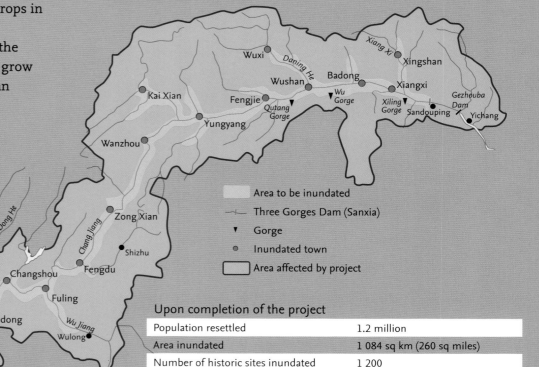

Area to be inundated
— Three Gorges Dam (Sanxia)
▼ Gorge
● Inundated town
☐ Area affected by project

Upon completion of the project

Population resettled	1.2 million
Area inundated	1 084 sq km (260 sq miles)
Number of historic sites inundated	1 200
Average installed capacity of HEP	84.7 billion kW per year
Money invested	180 billion yuan (22.5 billion $US)

Irrigated share of total cropland 1965–2002

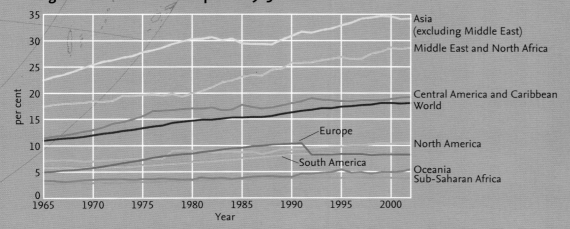

Asia (excluding Middle East)
Middle East and North Africa
Central America and Caribbean
World
Europe
North America
South America
Oceania
Sub-Saharan Africa

Today, dams have left big rivers like the Indus in Pakistan and the Yellow River in China regularly running dry. And many farmers are turning to underground water. But while the rains eventually replenish river flows, many underground water reserves are being emptied forever. Water tables are falling fast as countries such as Saudi Arabia and India pump up far more water than is replenished by rains. The world is growing ever more of its food with water which is not being replaced.

Euphrates river, Turkey 2 September 1976

The Southeastern Anatolia Project was planned in the 1970s to provide a means of supplying power and irrigation to this rural part of eastern Turkey. Prior to the construction of the Atatürk Dam on the Euphrates river this was an arid and agriculturally poor region, as illustrated by the brownness and small number of fields evident in the satellite image above.

By 1999 the landscape has been transformed by the construction and filling of the dam. In this image, large areas of green show irrigated farmland, especially around the town of Harran at the bottom right. Crop yields have increased and a commercial fishing industry is growing. However, as the project extracts water from the Euphrates there is concern from Syria and Iraq that not enough water is being released to meet their own needs downstream.

In 1973 the Mesopotamian marshlands, at the confluence of the Tigris and Euphrates rivers, were one of the world's great wetlands and home to the Marsh Arabs. Following the construction of the Atatürk (see previous pages) and other dams upstream, and the draining of the marshlands by the Iraqi regime in the 1990s, the whole ecosystem and way of life of the native people were affected. These false-colour satellite images highlight the changes which have occurred.

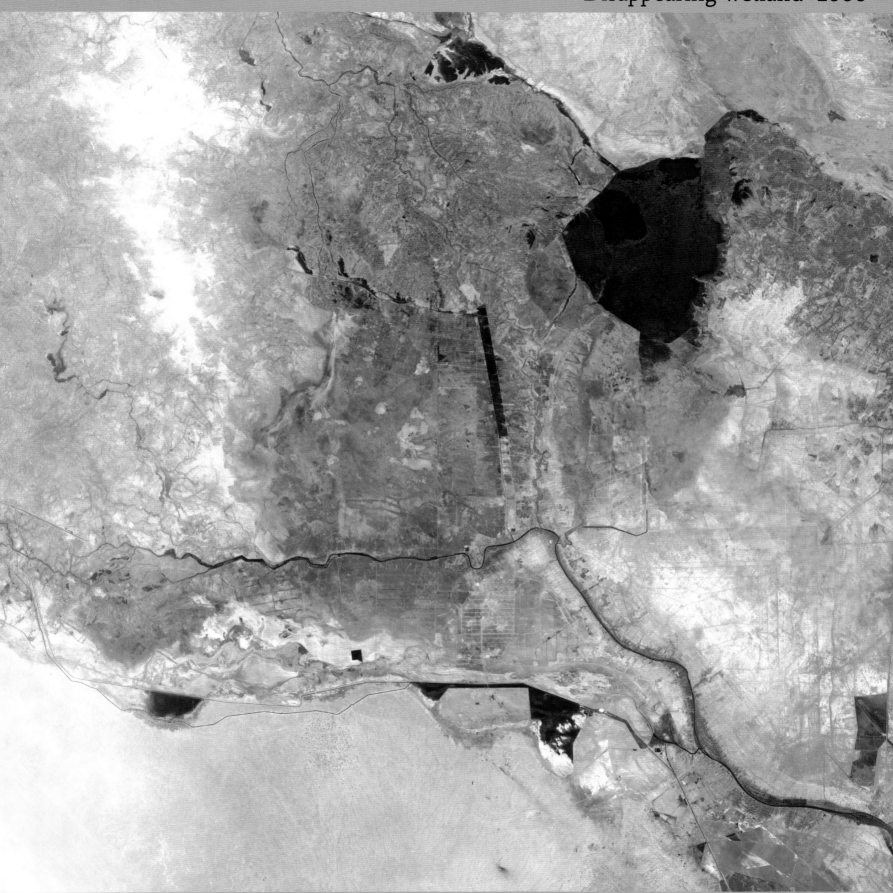

Water appears blue and the wetlands dark burgundy with farmland a brighter red. The difference between the two images is dramatic, with a vast area of marshland having disappeared to be replaced with dry barren land. However, since the change in the political situation in Iraq work has begun to restore the marshes, but in many areas the water now coming into the marshes is so polluted or salty that it may take many years for the area to return to anything like its former state.

Originally known as the Boulder Dam, the multi-purpose Hoover Dam was constructed in the 1930s to control flooding on the Colorado River, provide irrigation in an arid region, supply Los Angeles with water and generate electricity. Constructing the dam required the removal of 1 150 000 cubic m (1 500 000 cubic yards) of rock and soil. Completed in 1936, the dam holds back the 180 km-(110 mile)-long Lake Mead, the largest man-made reservoir in the USA. With the Art-Deco design of the dam and the recreation facilities offered by the lake the area attracts up to 10 million visitors each year.

Three Gorges Dam, China 21 February 2004 and 9 September 2004

The Three Gorges Dam project, started in 1993 and scheduled for completion by 2009, will raise the level of the Yangtze river behind the Sanxia Dam, which is already complete, by 110 m (361 feet). In doing so a reservoir 660 km (410 miles) long is being created, submerging 632 sq km (244 sq miles) of land, and displacing Gover one million people. In the upper photograph the marker shows the proposed lake level in Wushan by late 2006. By 2009 all the buildings below the new (white) ones will be submerged.

Wādī as-Sirḥān, Saudi Arabia   2 February 1986

In common with much of the Arabian Peninsula, Wādī as-Sirḥān in the Arabian Desert was an arid, barren region which struggled to support towns such as Tubarjal (highlighted above). Using oil revenues, Saudi Arabia introduced a centre pivot irrigation system to enable crops to be grown and to transform the landscape.

By 2004 the effect of the centre pivot system is seen in the above image. The desert is blooming, with each green dot a field irrigated by a rotating arm which distributes water raised by pump from an ancient aquifer below the ground. As a result, food production has increased and the future of the town is assured.

The World's major cities

Moscow
Los Angeles
New York
Mexico
Cairo
Karachi
Delhi
Beijing
Tōkyō
Ōsaka
Shanghai
Dhaka
Mumbai
Kolkata
Manila
Lagos
Jakarta
São Paulo
Rio de Janeiro
Buenos Aires

- ○ 5 million – 10 million
- ◯ 10 million – 20 million
- ⬤ over 20 million

Total urban population of major regions 1950–2030

World

Less developed regions[2]

Asia

More developed regions[1]
Africa
Latin America and the Caribbean[4]
Europe[3]
North America
Oceania

Population (millions)

5000

4000

3000

2000

1000

0

1950 1960 1970 1980 1990 2000 2010 2020 2030

Year

1. Europe, North America, Australia, New Zealand and Japan.
2. Africa, Asia (excluding Japan), Latin America and the Caribbean, and Oceania (excluding Australia and New Zealand).
3. Includes Russian Federation.
4. South America, Central America (including Mexico) and all Caribbean Islands.

Expanding cities

World's largest urban agglomerations
1955

Agglomeration	Country	Population (millions)
Tōkyō	Japan	13.713
New York	USA	13.219
London	U.K.	8.278
Shanghai	China	6.865
Paris	France	6.277
Buenos Aires	Argentina	5.843
Rhein-Ruhr	Germany	5.823
Moscow	Russian Federation	5.749
Chicago	USA	5.565
Los Angeles	USA	5.154

2005

Agglomeration	Country	Population (millions)
Tōkyō	Japan	35.327
Mexico City	Mexico	19.013
New York	USA	18.498
Mumbai (Bombay)	India	18.336
São Paulo	Brazil	18.333
Delhi	India	15.334
Kolkata	India	14.299
Buenos Aires	Argentina	13.349
Jakarta	Indonesia	13.194
Shanghai	China	12.665

2015 (projected)

Agglomeration	Country	Population (millions)
Tōkyō	Japan	36.214
Mumbai (Bombay)	India	22.645
Delhi	India	20.946
Mexico City	Mexico	20.647
São Paulo	Brazil	19.963
New York	USA	19.717
Dhaka	Bangladesh	17.907
Jakarta	Indonesia	17.498
Lagos	Nigeria	17.036
Kolkata	India	16.798

Very soon humanity will become a predominantly urban species. Cities occupy just 2 per cent of the Earth's land surface, but by 2007 they will house half of the world's population. And, perhaps most alarming of all, they use three-quarters of the resources we take from the Earth. The growth of the world's largest cities in particular has been staggering. The first megacity – with a population of ten million people – was New York, which reached that figure around 1940. Today there are twenty megacities, over half of them in Asia, including three each in China and India. The largest, Tōkyō in Japan, has over thirty-five million inhabitants. Most urban growth is from migration. China, which has ninety cities with populations of more than a million people, has a 'floating population' of more than 100 million people who have moved to cities in recent years. Its cities expect to welcome another 400 million people from the countryside within the next thirty years.

Cities have become economic powerhouses, providing jobs in manufacturing, transport and wholesaling, as well as services including universities, hospitals, government services, banking, media and culture. São Paulo contributes 40 per cent of Brazil's GDP; the Shanghai area takes a similar percentage of foreign investment in China. But, despite their attractions, cities can be centres of squalor too. More than a quarter of all urban inhabitants live in unplanned, overcrowded and often illegal squatter settlements, with no running water, let alone tickets to the opera. And cities can become victims of their size. Congestion, worsening air pollution and crime can cause megacities to stop growing, as people and businesses flee to the suburbs or surrounding new cities. Smoggy Mexico City had sixteen million people in 1984 and was widely expected to grow to thirty million by 2000; but instead it has stopped growing at less than twenty million. The result of this out-migration has been a new geographical phenomenon known as the 'polycentric megacity zone' – an urban landscape composed of a number of different centres. These new zones include the Yangtze delta region around Shanghai, southeast England around London, and the Japanese urban corridor between Tōkyō and Osaka. Better electronic communications are encouraging this trend.

The growth of cities with over 5 million inhabitants

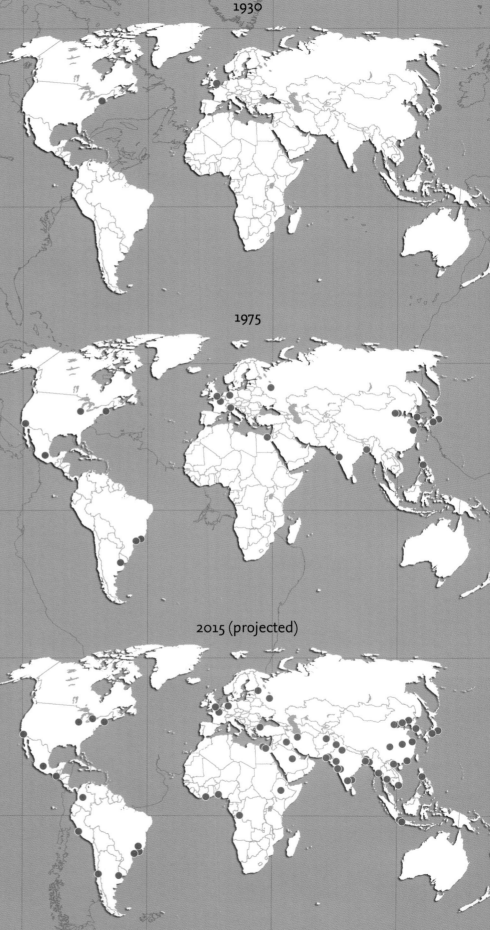

1930

1975

2015 (projected)

Urban sprawl itself is emerging as an environmental menace. Concrete and asphalt cover rich farmland around fast-expanding megacities such as the Indonesian capital Jakarta. And sprawl encourages ever greater use of motor vehicles, which cause smog and emit the gases behind climate change.

Cities have very large ecological 'footprints'. To service Londoners, and absorb the pollution they create, requires an area of land 120 times the size of the city itself. Sprawling cities such as Las Vegas have even bigger footprints. Many of the resources cities require may come from distant lands, but cities still put huge strains on local resources such as water. Air pollution from Chinese cities cuts crop yields across the country by up to one-third.

Yet some say that cities may be the key to a more sustainable future. In developing countries, people who move to cities from the countryside have fewer children. Urban children can be seen as a burden, requiring education for instance, rather than as a useful source of farm labour. So urbanization may be the ultimate solution to population growth.

Level of urbanization by major region 1970–2030

Urban population as a percentage of total population

	1975	2000	2030
World	37.3	48.3	60.8
More developed regions[1]	67.2	74.5	81.7
Less developed regions[2]	26.9	42.1	57.1
Africa	25.3	38.7	53.5
Asia	24.0	38.8	54.5
Europe[3]	66.0	73.0	79.6
Latin America and the Caribbean[4]	61.2	76.8	84.6
North America	73.8	80.2	86.9
Oceania	71.7	73.1	74.9

1. Europe, North America, Australia, New Zealand and Japan.
2. Africa, Asia (excluding Japan), Latin America and the Caribbean, and Oceania (excluding Australia and New Zealand).
3. Includes Russian Federation.
4. South America, Central America (including Mexico) and all Caribbean Islands.

The inhabitants of cities recycle more of their waste, and use public transport more often than their country cousins. They even grow food. The proximity of millions of customers means every scrap of spare land is cultivated. Around 15 per cent of the world's food is grown within city limits. Shanghai still produces most of its own milk, eggs and vegetables. Amazingly too, cities can be centres of wildlife. One small derelict industrial site in London has 300 species of plants growing – many times more than the countryside around the city. Cities may still be global parasites but they have redeeming features.

Tallest buildings 1975 and 2006

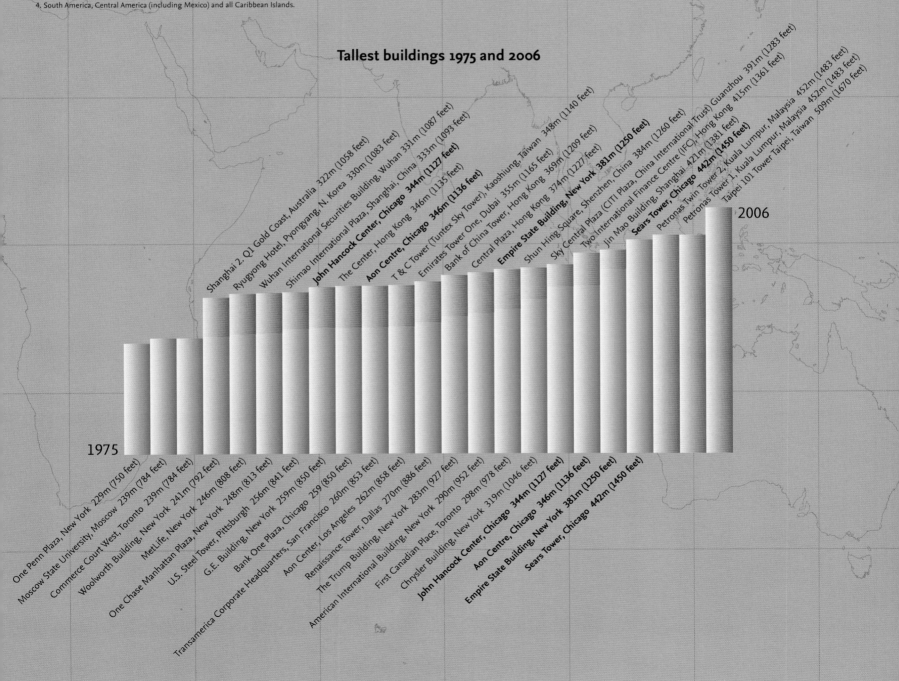

1975

2006

One Penn Plaza, New York 229m (750 feet)
Moscow State University, Moscow 239m (784 feet)
Commerce Court West, Toronto 239m (784 feet)
Woolworth Building, New York 241m (792 feet)
MetLife, New York 246m (808 feet)
One Chase Manhattan Plaza, New York 248m (813 feet)
U.S. Steel Tower, Pittsburgh 256m (841 feet)
G.E. Building, New York 259m (850 feet)
Bank One Plaza, Chicago 259m (850 feet)
Transamerica Corporate Headquarters, San Francisco 260m (853 feet)
Aon Center, Los Angeles 262m (858 feet)
Renaissance Tower, Dallas 270m (886 feet)
The Trump Building, New York 283m (927 feet)
American International Building, New York 290m (952 feet)
First Canadian Place, Toronto 298m (978 feet)
Chrysler Building, New York 319m (1046 feet)
John Hancock Center, Chicago 344m (1127 feet)
Aon Centre, Chicago 346m (1136 feet)
Empire State Building, New York 381m (1250 feet)
Sears Tower, Chicago 442m (1450 feet)

Shanghai 2. Q1 Gold Coast, Australia 322m (1058 feet)
Ryugyong Hotel, Pyongyang, N. Korea 330m (1083 feet)
Wuhan International Securities Building, Wuhan 331m (1087 feet)
Shimao International Plaza, Shanghai, China 333m (1093 feet)
John Hancock Center, Chicago 344m (1127 feet)
The Center, Hong Kong 346m (1135 feet)
Aon Centre, Chicago 346m (1136 feet)
T & C Tower (Tuntex Sky Tower), Kaohsiung, Taiwan 348m (1140 feet)
Emirates Tower One, Dubai 355m (1165 feet)
Bank of China Tower, Hong Kong 369m (1209 feet)
Central Plaza, Hong Kong 374m (1227 feet)
Empire State Building, New York 381m (1250 feet)
Shun Hing Square, Shenzhen, China 384m (1260 feet)
Sky Central Plaza (CITI Plaza, China International Trust) Guanzhou 391m (1283 feet)
Two International Finance Centre (IFC), Hong Kong 415m (1361 feet)
Jin Mao Building, Shanghai 421m (1381 feet)
Sears Tower, Chicago 442m (1450 feet)
Petronas Twin Tower 2, Kuala Lumpur, Malaysia 452m (1483 feet)
Petronas Tower 1, Kuala Lumpur, Malaysia 452m (1483 feet)
Taipei 101 Tower Taipei, Taiwan 509m (1670 feet)

Aberdeen Harbour, Hong Kong 1920

Taken in 1920, this photograph of Aberdeen harbour, Hong Kong shows traditional junks at anchor in the harbour. The surrounding land contains few buildings although in the distance some 'low rise' buildings are visible. Contrast that with the recent view opposite.

The entire harbour area is now enclosed by multi-storey apartment blocks built to accommodate the ever-increasing population. When land is at a premium, after reclaiming some from the sea the only way is up.

When this photograph of Singapore was taken, the multi-storey Asia Insurance building to the left was the tallest in the city and the harbour was used by traditional sampans and rowing boats.

Although the Asia Insurance building still exists, it is no longer the tallest in the city. It is now dwarfed by many multi-storey commercial buildings which have sprung up in the last forty years in the central area of the city as Singapore has expanded as a major commercial centre. Even away from the business district much of the land of Singapore island is built up with the majority of Singaporeans living in high-rise flats.

Las Vegas, USA 13 May 1973

Although this satellite image taken in 1973 is at a low resolution, the combination of the grid-iron street pattern and the green of golf courses, parks and gardens illustrates that even at this time Las Vegas was an expanding city.

t is now the fastest-growing metropolitan area in the USA and in this 2000 satellite image the desert has turned green in many areas as suburban growth heads outwards from the city centre. The city's expansion has been in all directions and the new roads bypassing the city to

Kathmandu, Nepal 1969 and today

In the thirty-year period from 1969, when the top photograph was taken, the population of Kathmandu, the capital of Nepal, increased five-fold due mostly to migration from the countryside. This resulted in the built-up area of the city expanding outwards to swallow up rural communities. The lower photograph, taken from the same location, shows a dramatic change in the landscape in this relatively short time.

The shanty town of Favela Morumbi is one of São Paulo's largest slum areas and is typical of the unplanned expansion seen in many underdeveloped or developing countries as thousands of country dwellers flood into the main cites in search of a better life. Conditions are often poor, with little or no sanitation and in some cities such unofficial developments are razed to the ground by the authorities, only to spring up again.

Travel and transportation

Since his original exit from Africa, mankind has always been a traveller. Modern jet-setters have much in common with their nomadic ancestors. But today's travellers, rather than living off the land, remake the landscape as they go.

In most cities today, as much land is given over to paved highways as to buildings themselves. In most countries, the natural landscape, including drainage networks and the migration paths of animals, is fragmented by roads and railways and pipelines. Airport hubs such as London's Heathrow, the world's largest international airport, can be as big as a medium-sized city.

International business and leisure travel now represents, by some estimates, 10 per cent of the entire global economy. Transport is the biggest single source of greenhouse gases. Civil aircraft alone contribute 5 per cent of total emissions, and air traffic is expected to double by 2020.

Busiest scheduled international air passenger routes 2000

Route	Annual number of passengers (both directions)
London–New York	3 832 630
London–Amsterdam	3 334 541
London–Paris	2 655 116
Hong Kong–T'aipei	2 617 471
Seoul–Tōkyō	2 314 950
Kuala Lumpur–Singapore	2 166 985
Hong Kong–Bangkok	2 158 923
London–Dublin	2 019 221
Bangkok–Singapore	2 014 392
London–Frankfurt	1 943 328
Hong Kong–Tōkyō	1 813 558
Honolulu–Tōkyō	1 796 052
Hong Kong–Singapore	1 592 101
Hong Kong–Manila	1 557 552
New York–Paris	1 537 225
Jakarta–Singapore	1 499 975
London–Madrid	1 408 868
Madrid–Paris	1 378 871
Ōsaka–Seoul	1 362 012
New York–Toronto	1 342 454
London–Los Angeles	1 318 627
Tōkyō–Bangkok	1 317 245
Hong Kong–Seoul	1 307 414
London–Chicago	1 270 937
Singapore–Tōkyō	1 230 769
London–Brussels	1 193 121
Frankfurt–New York	1 182 741
Tōkyō–Los Angeles	1 114 101
London–Boston	1 109 692
London–Barcelona	1 105 671
London–Zurich	1 100 064
Frankfurt–Paris	1 078 592
London–Stockholm	1 072 624
London–Munich	1 064 205
Tōkyō–T'aipei	1 058 265
Chicago–Toronto	1 033 936
London–San Francisco	1 012 837
London–Nice	971 248
London–Washington	968 510
Paris–Amsterdam	966 565

Road and railway growth 1990 and 2005 (selected countries)

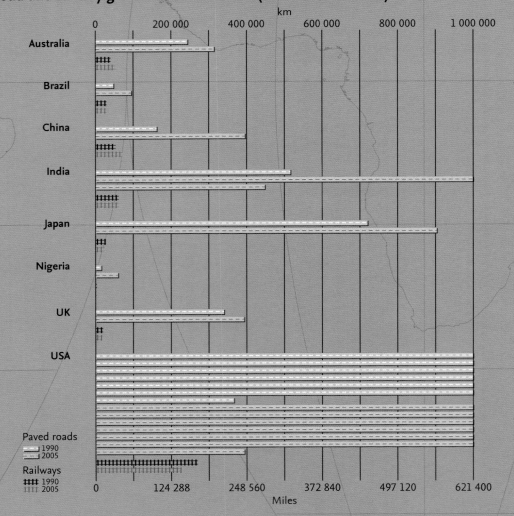

Countries: Australia, Brazil, China, India, Japan, Nigeria, UK, USA

Paved roads
— 1990
— 2005

Railways
╫╫╫ 1990
╫╫╫ 2005

Airport growth 1990 and 2005 (selected countries)

Airports with permanent surfaced runway

Country	1990	2005
Australia	235	305
Brazil	386	698
China	260	383
India	202	234
Japan	128	143
Nigeria	32	36
United Kingdom	245	334
United States of America	4 820	5 128

Busiest airports 1977 and 2005

International and domestic flights

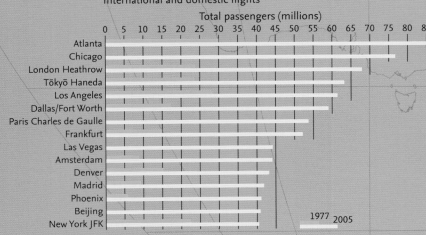

New transport connections are spanning the seas – connecting Denmark to Sweden by bridge, and England and France by tunnel. And they are 'opening up' to migrants and economic development the remaining thinly populated parts of the planet. China's railway to Lhasa, the capital of Tibet, and Brazil's new roads into the heart of the Amazon rainforest are but two examples.

High-speed railway links are bringing European capitals such as London, Paris and Brussels closer together. By the end of the decade China will have completed a magnetic levitation (Maglev) railway from Shanghai to Hangzhou, a journey of 170 km (106 miles) which will take just twenty-seven minutes.

Meanwhile, cheap air travel is changing cultural geography, making London partygoers familiar with Latvia, and taking Japanese businessmen on golfing weekends to Cambodia. Thanks to international travel opportunities, an estimated 120 million people are currently living and working outside their countries of origin.

The ecological impacts can be great. Mass tourism has destroyed Mediterranean freshwater reserves, Himalayan forests and Caribbean coral reefs – but also brought funds for conserving Kenyan elephants and Costa Rican jungles. In future it may upset the ecological integrity of Antarctica.

The steep-sided Tarn river valley near Millau, southern France, was for a long time an obstacle to the free flow of traffic between the north and south of the country. The descent into the valley and past the town of Millau resulted in traffic bottlenecks, especially in the main tourist season.

Construction of a bridge to remove the traffic problem began in 2001 after various options were considered. This new bridge was completed and opened to traffic in December 2004. It is 2.46 km (1.53 miles) long and its highest pier is 336 m (1102 feet) tall, making it the tallest vehicular bridge in the world.

Gravelly Hill, England, UK 4 June 1961

This black and white aerial photograph was taken two years before the construction of the Gravelly Hill Interchange. The road junction was to be built to provide access from the M6 motorway from London to the centre of Birmingham, the UK's second largest city, and to link with a network of local roads.

The interchange soon became known as 'Spaghetti Junction' owing to its complexity of roads. At the heart of England and a transport focal point the site also features canals and railways built in the nineteenth and early twentieth centuries to speed the carriage of industrial

Öresund bridge and tunnel, Denmark/Sweden

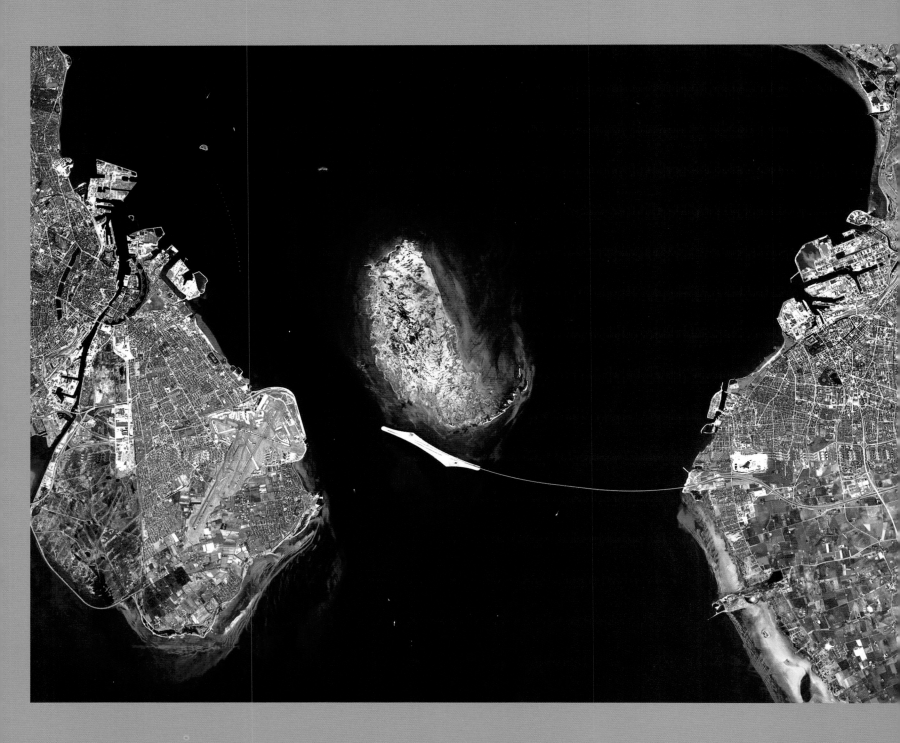

The Öresund bridge and tunnel combines both road and rail transport in a link which extends for 15.5 km (9.6 miles) between Denmark and Sweden. The bridge is 7.8 km (4.8 miles) long and the tunnel 3.5 km (2.2 miles). The rest of the distance is taken up by Peberholm, an artificial island built in the Öresund strait. In this satellite image the associated network of roads can be seen linking both ends of the crossing. When it was opened in 2004 it completed a road and rail link from Sweden across Denmark to the North Sea port of Esbjerg on Denmark's west coast.

The Qingzang railway, between Lhasa and Golmud in China, is the highest railway in the world. Completed in 2006, the line connects Xining, Qinghai Province with Lhasa, Tibet. Where the railway crosses the Tanggula Pass at 5072 m (16 640 feet) it is the highest railway in the world. As a result of the high altitude, special passenger carriages are used which are equipped with enriched oxygen and UV-protection systems.

Damaged

World

Man's impact

Deforestation – *the felling of trees, often illegally and in an unmanaged way, commonly to provide land for agriculture or industry*

Conflict – *disputes between countries or groups of people based on territorial claims, ideology, religion, ethnicity or competition for resources*

Pollution – *the release into the environment of harmful or poisonous substances, often by-products of industrial or agricultural activity*

Deforestation

Forests are the planet's largest, most biologically diverse and most critical ecosystems. They once covered some two-thirds of the Earth's land area, and still cover approximately a third, nearly 40 million sq km (15.4 million sq miles). They stretch from the tropics to the edges of the Arctic tundra, and contain around half the world's plant and animal species. They play vital roles in recycling water around the planet, protecting soils and maintaining the natural carbon cycle. Mankind has always lived in, planted and destroyed forests. There are probably few true virgin forests anywhere. But the scale and pace of deforestation in the past 200 years dwarfs anything seen before. Some 99 per cent of Europe's 'old growth' forests outside Russian Federation are gone, along with 95 per cent of those in the continental USA. They have been either converted to farmland or urban areas or have been replaced with commercial plantations. Larger stands remain in the 'boreal' forests of northern Russian Federation, Canada and Alaska, and in the equatorial rainforests of the Amazon, central Africa and southeast Asia. But many of these forests are now disappearing fast, and could be gone within a decade.

Distribution of forests by region 2005

Region	Forest area sq km	Forest area sq miles	% of global forest area	% of region's land area
World	39 520 260	15 258 773	100.0	30.3
Africa	6 354 120	2 453 326	16.1	21.4
Asia	5 715 770	2 206 859	14.5	18.5
Europe	10 013 940	3 866 382	25.3	44.3
North and Central America	7 058 490	2 725 283	17.9	32.9
Oceania	2 062 540	796 347	5.2	24.3
South America	8 315 400	3 210 576	21.0	47.7

Annual net change in forest area by region, 1990–2005

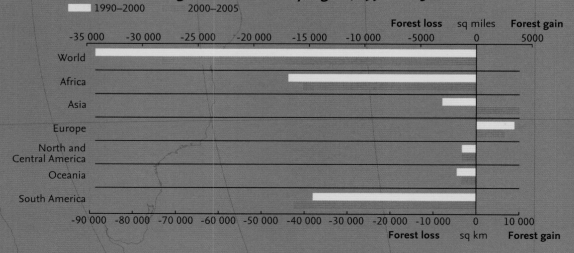

1990–2000   2000–2005

Forest loss   sq miles   Forest gain

-35 000   -30 000   -25 000   -20 000   -15 000   -10 000   -5000   0   5000

World
Africa
Asia
Europe
North and Central America
Oceania
South America

-90 000   -80 000   -70 000   -60 000   -50 000   -40 000   -30 000   -20 000   -10 000   0   10 000

Forest loss   sq km   Forest gain

The world's forests

Evergreen broadleaf forest
Evergreen needleleaf forest
Deciduous needleleaf forest
Deciduous broadleaf forest
Mixed forest

The pressures are greater in more densely populated countries, where migrant farmers burn forests to provide land for farming. But in the Amazon Basin, landowners are clearing forests to make way for cattle ranching and soya bean production, while logging is escalating in Africa. Indonesian islands such as Borneo and Sumatra suffer from both logging and clearance for palm oil plantations. Ironically, one growing market for palm oil is as an environmentally friendly 'biofuel' additive to vehicle fuels in Europe. The scale of deforestation is not the only issue, however. Piecemeal forest removal and penetration by roads break large forests into fragments which are less able to support a full range of species, because large animals have less room to roam and because fragmentation limits an ecosystem's ability to recover from catastrophes such as fires.

Lawlessness is a major problem in large, remote forests. The rights of indigenous communities to manage the forests are often abused. And conservation laws are widely flouted. An estimated 80 per cent of all logging in Indonesia is now illegal. In poor countries where forests are one of the few valuable natural resources, corruption in government and money-raising by armies, rebel groups and militias often contribute to forest destruction. Liberia, Myanmar (Burma) and Papua New Guinea are recent well-documented examples. Following a decision in 1998 to end logging of natural forests within its own borders, China is now the world's largest importer of tropical rainforest timber. But the timber, much of it illegally cut, ends up in products such as furniture which are sold in Europe, Japan and North America.

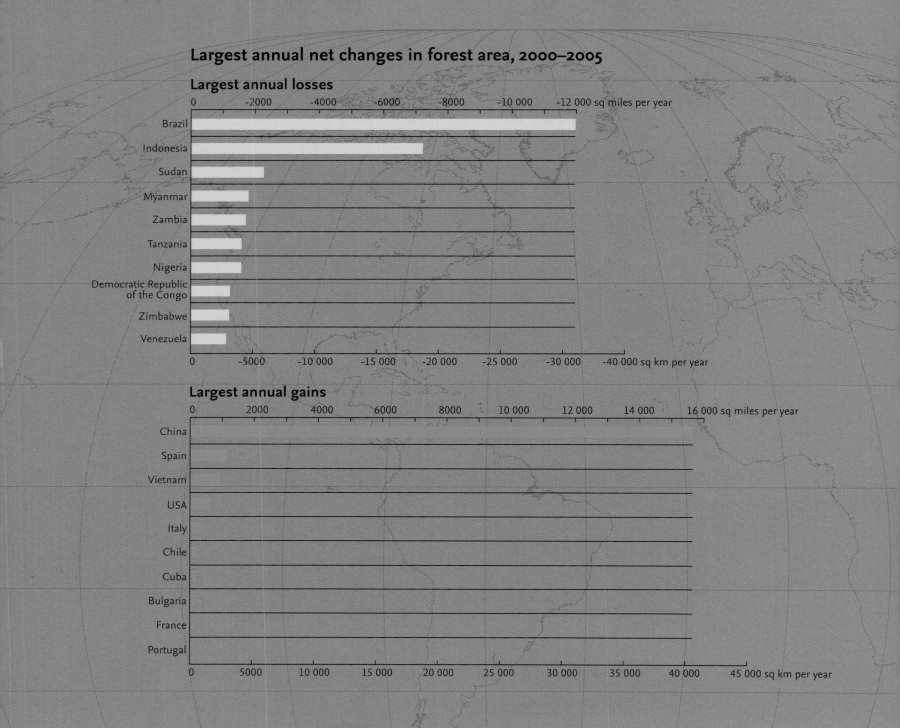

Largest annual net changes in forest area, 2000–2005

Largest annual losses

Deforestation in the Amazon Basin, Brazil

- Areas of deforestation
- Tropical forest
- Other vegetation

BRAZIL

VENEZUELA

GUYANA SURINAME **French Guiana**

COLOMBIA

Macapá

Marajó Island

Belém

Manaus Santarém São Luís

A M A Z O N I A

Maraba

Branco

Amazon

Amazon

Juruá

Madeira

Tapajós

Purus

Xingu

Araguaia

Tocantins

Porto Velho

PERU

Rio Branco

BOLIVIA

Cuiabá

B R A Z I L

500 km
311 miles

Countries with largest forest area
Percentage of total world forest area

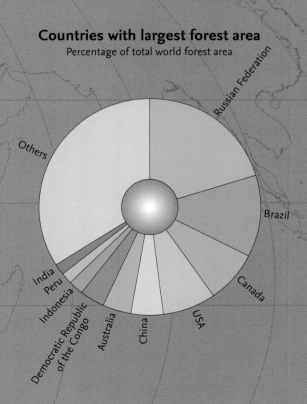

Others

Russian Federation

Brazil

Canada

USA

China

Australia

Democratic Republic of the Congo

Indonesia

Peru

India

When forests are removed, soils often erode rapidly when exposed for the first time to direct rain. But in the longer run the destruction of forests may diminish rainfall. This is because, particularly in lush rainforests, trees collect and recycle rainfall back into the air through a process called evapotranspiration, keeping the winds moist and stimulating rainfall downwind. Remove the trees and the rains fail. The loss of rainforests in West Africa is thought to be one reason for the spreading Sahara desert. Deforestation can also alter global climates indirectly. Decaying timber eventually releases into the atmosphere carbon dioxide, a greenhouse gas which accelerates global warming. The loss of 1 sq km (0.4 sq miles) of forest will typically release 10 000 tonnes of carbon. Up to 20 per cent of current global warming may be due to carbon released from deforestation. In the past two decades, the temperate forests of Europe and North America have undergone modest increases in extent, thanks to the planting of commercial stands. Across the world, more forests are being declared national parks and reserves, but conservationists say many of these 'paper parks' offer little protection to surviving natural forests.

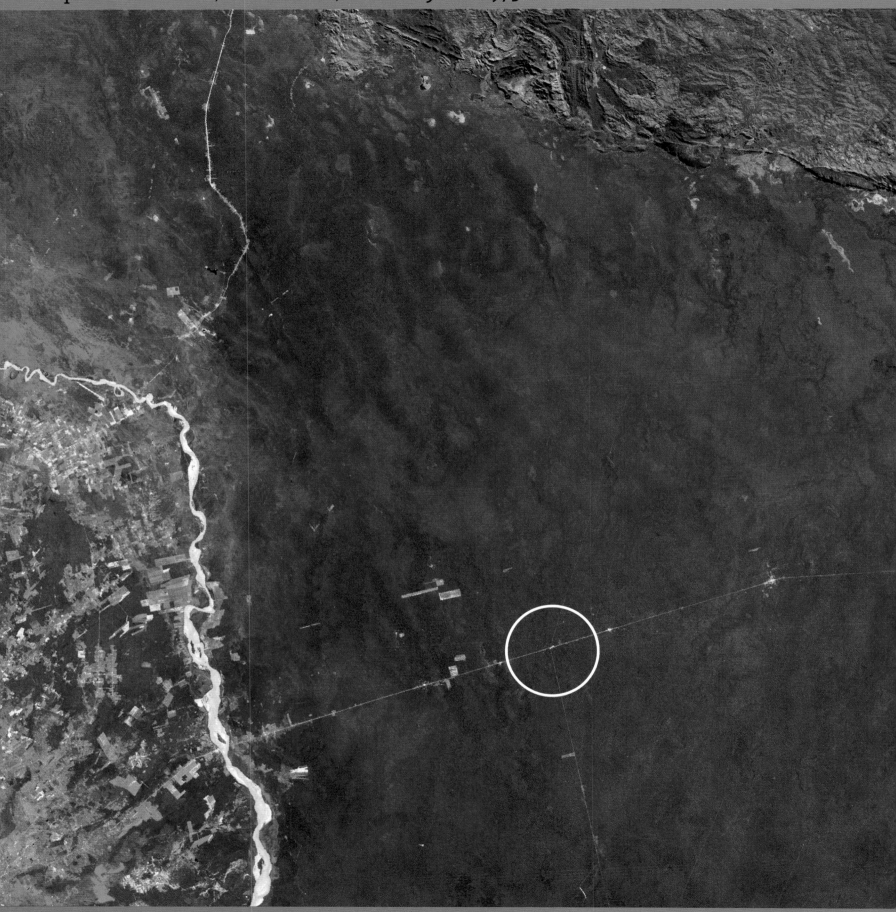

Tropical rainforest, Santa Cruz, Bolivia June 1975

In 1975, the area adjacent to the Rio Grande ó Guapay River, northeast of the Bolivian city of Santa Cruz, was one of rich, dense rainforest crossed by a handful of tracks. The whole region was sparsely populated with only occasional forest clearings visible in this satellite image. Increased agricultural activity is evident in the lower left corner, towards the city.

By 2003, after a period of extensive deforestation, the area had been transformed into a major agricultural area, mainly growing soya beans for export. The population of Santa Cruz has increased from less than 30 000 to over 1 million in the last 35 years. This has put enormous pressure on the forest with an increasing demand for new settlements and farmland.

Some of the most spectacular waterfalls in the world, the Iguaçu Falls, lie at the junction of Argentina, Brazil and Paraguay. The isolation of this region gave rise to the unique Paranaense rain forest ecosystem which supported thousands of species unique to the region. This satellite image shows early evidence of deforestation, with patterns of tree felling following lines of communication.

Over a period of thirty years the changes have been dramatic. Vast areas of forest have been cleared for agriculture, particularly in Paraguay on the left of the image. This process was accelerated by the creation of a huge new reservoir, following the construction of the Itaipu Dam (circled) on the Paraná River in the early 1980s. Some forest on the right of the image lies within Iguaçu National Park and has been protected from destruction.

The contrast between native tropical rainforest and agricultural land use developed after deforestation is clear in this image of the Iguaçu river in Brazil. Across the world, huge areas of forest are being cleared every year for logs, livestock or crops. World demand for soya beans in particular has led to much of this clearance. Forest, and its rich and unique diversity of wildlife, can remain untouched if it is protected for

conservation within national parks and reserves, or if it is not yet economical to clear it. In this case, the near bank of the river is part of Iguaçu National Park and is protected from deforestation.

The Amazon Basin is the largest drainage basin in the world, covering over 7 million sq km (2.7 million sq miles) of which approximately 70 per cent is rainforest. It is the wettest region in the world with an average annual rainfall of over 2.5 m (8.2 ft). The rainforest supports 30 per cent of all known plant and animal species, many of which are unique to the region.

Forest cleared for cattle, Rio Branco, Brazil

Burning is a commonly used method of clearing forest for agricultural use. It totally destroys the fragile ecosystem and creates land which remains fertile for only a few years. Over 30 000 sq km (11 500 sq miles) of forest – an area the size of Belgium – are cleared each year in Brazil alone. This rate of deforestation has an immediate impact on local biodiversity and on the global environment.

World conflict and military spending

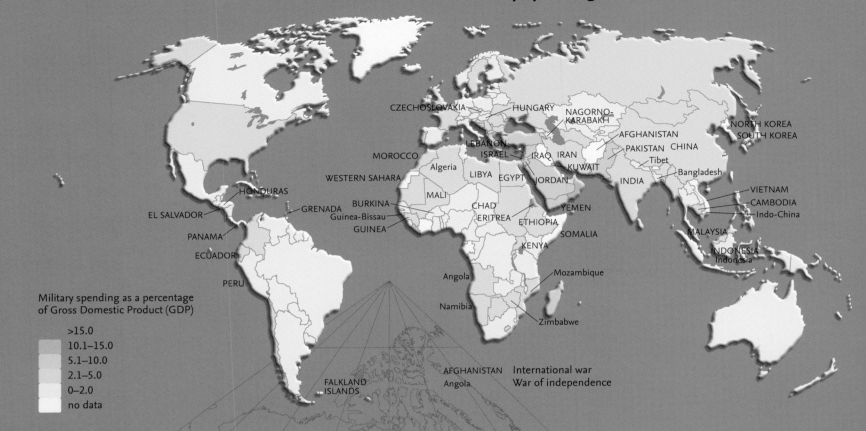

Military spending as a percentage
of Gross Domestic Product (GDP)

- >15.0
- 10.1–15.0
- 5.1–10.0
- 2.1–5.0
- 0–2.0
- no data

AFGHANISTAN   International war
Angola   War of independence

Conflict

Since the earliest city states of Mesopotamia 6000 years ago, disputes not only over land and borders, but also over resources such as water, forests and farmland have triggered conflicts. Rivers crossing international borders have been a particular bone of contention. Competition for the water of the River Jordan was one factor in the 1967 war between Israel and its neighbours, and international rivers have caused conflicts between India and Pakistan, Turkey and Syria, and Egypt and Ethiopia. Conflicts can scar the landscape, either deliberately through the 'scorched earth' tactics of generals, or accidentally. When China dynamited the levees on the Yellow River in 1938 to halt advancing Japanese soldiers, the ensuing floods killed almost a million people, and it took a decade to divert the river back to its old course. Defoliants destroyed large areas of rainforest in Cambodia and Vietnam during the Indochina war in the 1960s. After the first Gulf War, Saddam Hussein sent engineers to drain the Mesopotamian marshes, the largest wetland in the Middle East, to flush out opponents. Recent conflicts and terrorist acts have broken irrigation networks across Iraq, caused toxic pollution from bombed factories in former Yugoslavia, deforested Afghanistan and Liberia, wiped out big game in central Africa, covered parts of Kuwait in oil from burning wells and filled the streets of Manhattan with toxic dust from the collapsed Twin Towers.

Terrorist Incidents by Region 1998–2005

	Incidents	Fatalities	Injuries
Middle East	4 355	5 385	12 814
Western Europe	2 672	336	1 463
South Asia	2 541	4 173	10 404
Latin America	1 461	1 341	2 095
Eastern Europe	1 035	1 808	4 664
Southeast Asia and Oceania	431	918	3 138
Africa	311	2 127	7 249
North America	103	2 994	30
East and Central Asia	83	134	151

Major terrorist incidents

	Date	Summary	Fatalities	Injuries
Lockerbie, Scotland	December 1988	Airline bombing	270	5
Tōkyō, Japan	March 1995	Sarin gas attack on subway	12	5 700
Oklahoma City, USA	April 1995	Bomb in the Federal building	168	over 500
Nairobi, Kenya and Dar es Salaam, Tanzania	August 1998	US Embassy bombings	257	over 4 000
Omagh, Northern Ireland	August 1998	Town centre bombing	29	330
New York and Washington D.C., USA	September 2001	Airline hijacking and crashing	2 752	4 300
Bali, Indonesia	October 2002	Car bomb outside nightclub	202	300
Moscow, Russian Federation	October 2002	Theatre siege	170	over 600
Bāghdad and Karbalā', Iraq	March 2004	Suicide bombing of pilgrims	181	over 400
Madrid, Spain	March 2004	Train bombings	191	1 800
Beslan, Russian Federation	September 2004	School siege	330	700
London, UK	July 2005	Underground and bus bombings	52	700
Sharm ash Shaykh, Egypt	July 2005	Bombs at tourist sites	88	200

Such disputes can result in huge numbers of refugees. Displaced and forced to live off the land, they in turn can often cause environmental damage. The millions who fled Rwanda during the 1990s survived by destroying large areas of protected forest in the neighbouring Democratic Republic of the Congo. The Palestinians in Gaza have exhausted local water reserves. Barriers designed to prevent conflicts or terrorist acts, such as Israel's barrier around the West Bank, have changed the landscape and have greatly affected the daily lives of people. Ironically, such barriers can sometimes be good for nature. The heavily armed border between East and West Germany during the Cold War created a green corridor for nature amid the minefields, and similar areas remain between North and South Korea and the Turkish and Greek halves of Cyprus.

Middle East politics

Changing boundaries in Israel/Palestine since 1922

West Bank

Security

18 per cent of land under Palestinian control
23 per cent of land under Palestinian civil control and joint security control
59 per cent of land under Israeli control

Population

97 per cent Palestinian Arab
687 500 refugees

Gaza

Security

100 per cent of land under Palestinian control since Israeli withdrawal 2005

Population

98 per cent Palestinian Arab
961 600 refugees

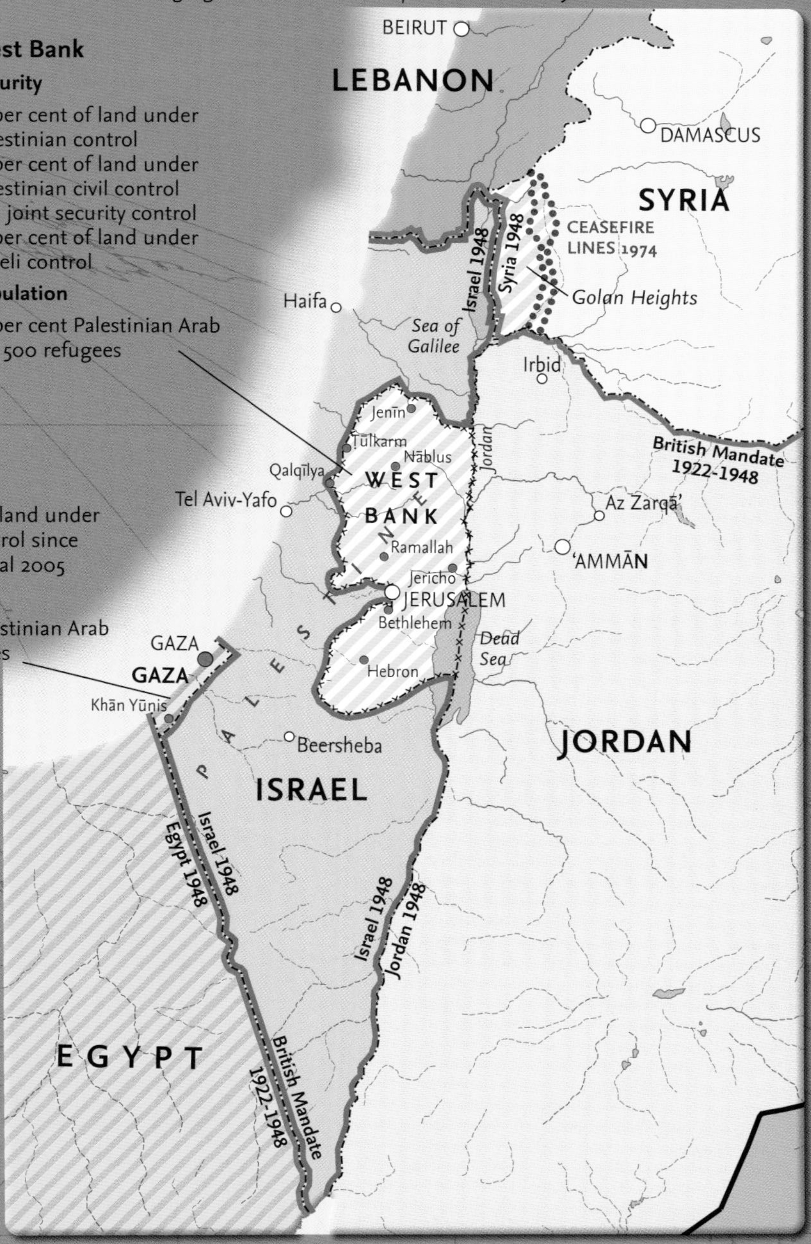
BEIRUT
LEBANON
DAMASCUS
SYRIA
CEASEFIRE LINES 1974
Golan Heights
Israel 1948
Syria 1948
Haifa
Sea of Galilee
Irbid
British Mandate 1922-1948
Jenīn
Ţūlkarm
Nāblus
Qalqīlya
WEST BANK
Tel Aviv-Yafo
Az Zarqā'
Ramallah
'AMMĀN
Jericho
JERUSALEM
Bethlehem
Jordan
Dead Sea
GAZA
GAZA
Khān Yūnis
Hebron
PALESTINE
Beersheba
ISRAEL
JORDAN
Israel 1948
Egypt 1948
Israel 1948
Jordan 1948
EGYPT
British Mandate 1922-1948

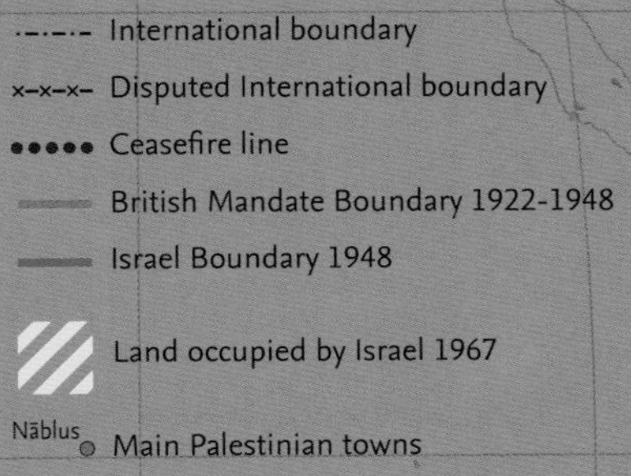
International boundary
Disputed International boundary
Ceasefire line
British Mandate Boundary 1922-1948
Israel Boundary 1948
Land occupied by Israel 1967
Nāblus
Main Palestinian towns

The devastation of an atomic bomb, Hiroshima, Japan   September 1945

At 8:15 am on 6 August 1945, the first atomic bomb to be used in warfare exploded over the Japanese city of Hiroshima. Over 80 000 people were instantly killed, and the total death toll rose to approximately 140 000. The only building left standing within a 3 km (1.9 mile) radius of the blast was the Prefectural Industrial Promotion Hall.

Preserved in its condition immediately after the bomb, and now designated as a World Heritage Site, the Hiroshima Peace Memorial serves as a symbol of the destructive force of nuclear weapons. Hiroshima today, a thriving city of over one million people, describes itself as a city of international peace and culture, and serves as testimony to the power of life and creation over destruction.

Kuwait during and after the Gulf war February 1991 and May 2001

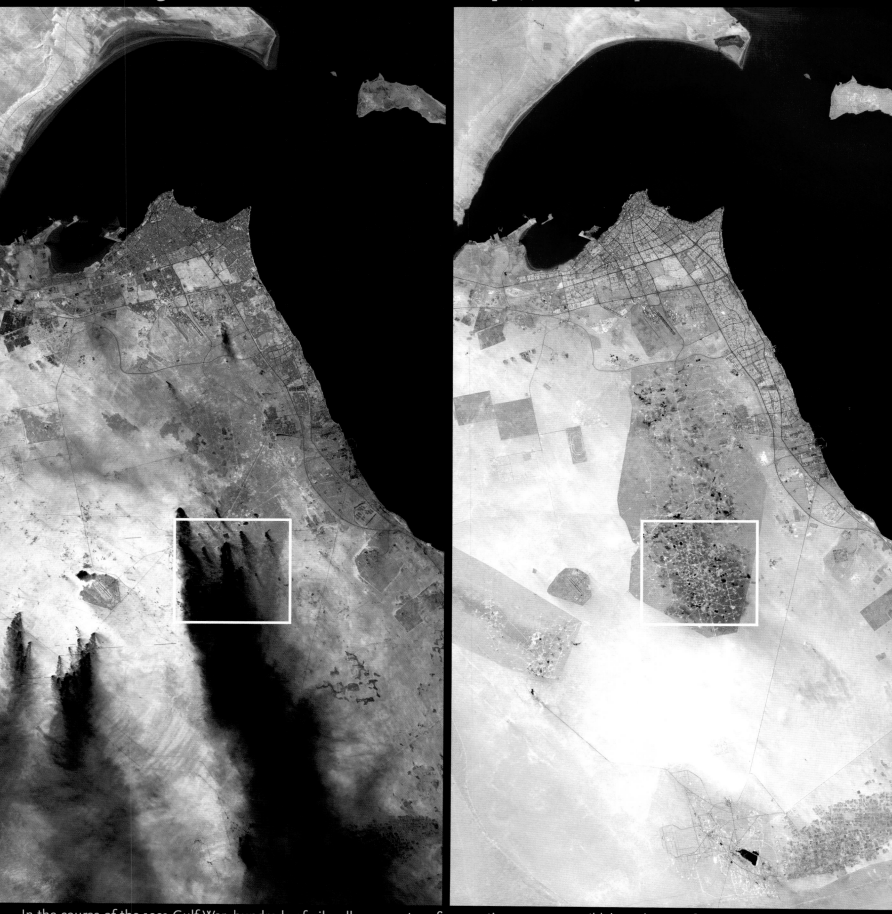

In the course of the 1991 Gulf War, hundreds of oil wells were set on fire, creating numerous oil lakes. The soot from the fires combined with sand and oil to leave a black layer of 'tarcrete' on almost five per cent of the country's area. This can be seen

The Bamiyan Valley in Afghanistan was an important pilgrimage site and Buddhist centre. Amongst its treasures were two stone Buddhas – at 55 m (180 feet) high, the tallest in the world. The Buddhas were totally destroyed by the ruling Taliban regime who believed that statues were un-Islamic. The site is now a World Heritage Site and the present Afghan government hopes that the statues can be restored.

The twin towers of the World Trade Center in New York were one target in the terrorist attacks on the USA by al-Qaeda on 11 September 2001. The towers were completely destroyed after being hit by passenger aircraft which flew directly into them. Over 2700 people were killed and more than 4000 injured in the attacks.

The skyline of New York was changed forever by the 9/11 attacks as these views south over Manhattan, towards the mouth of the Hudson River, show. The towers were symbols of the city's success and its position as a major world city and financial centre. Their destruction left a gaping hole in virtually any view of the city.

The Iraqi leader Saddam Hussein had several palaces within the Iraqi capital Baghdad, from which he controlled the country. This satellite image shows the Republican Palace (highlighted), close to the Presidential Palace and the Tigris river. When Saddam refused to comply with UN resolutions regarding disarmament, and military action against him became likely, these palaces were quickly identified as military targets.

In March 2003 US-led coalition forces began attacks on Iraq. A 'shock and awe' campaign of air strikes was centred on Baghdad. This image shows the extensive damage caused to the Republican Palace, which housed Saddam's command and control bunker. The surrounding area remains relatively unscathed, although a fire continues to burn to the west as a result of another missile strike.

Qalqilya, West Bank, Palestine March 2002 and June 2003

On the border between Israel and the West Bank, Qalqilya was known as the West Bank's 'fruit basket' because of the productivity of its surrounding farmland. In 2002, Israel began constructing a 'security barrier' around the West Bank to protect itself from suicide bombers contesting Israel's occupation. By June 2003 Qalqilya was surrounded on three sides by this barrier, cut off from the farms on which it depends.

The barrier being built around the West Bank is seen by Israel as essential to maintaining its security and in preventing the movement of Palestinian terrorists into Israel. However, the legality of the existence and route of the barrier is strongly contested. The daily lives of the Palestinian population are greatly affected, with access to schools, hospitals and farmland severely restricted.

CO2 emissions
Emissions of CO2 per person

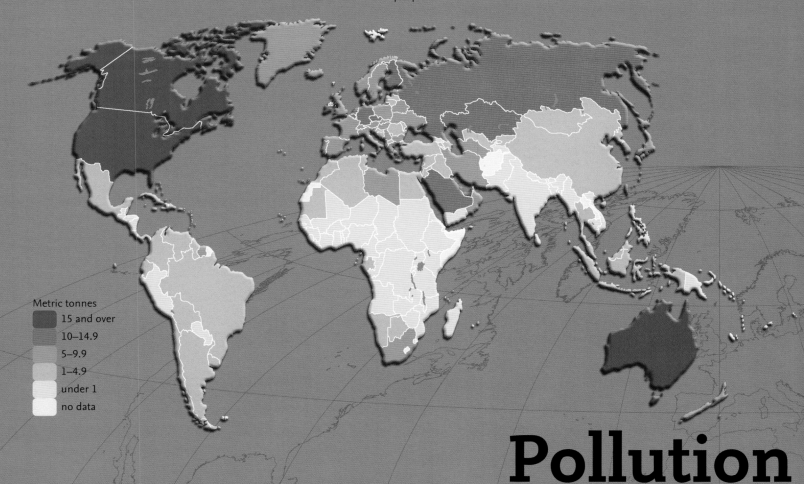

Metric tonnes
- 15 and over
- 10–14.9
- 5–9.9
- 1–4.9
- under 1
- no data

Pollution

Pollution, the effect of the discarded wastes of our consumer society, despoils the landscape at every scale from the small back yard to the oceans and the global atmosphere. Sometimes nature can make use of our waste, but mostly we discard it in places and in volumes which undermine natural systems and often damage human health. Air pollution, mostly from burning fuel but also from forest fires, blankets ever larger areas. Globally, 50 per cent of chronic respiratory disease is due to air pollution, notably caused by fine particles (less than 10 microns in diameter – PM_{10} particulates) which can penetrate deep into the respiratory tract. Millions die prematurely each year as a result. Urban smog creates acid rain which kills trees, damages crops and makes lakes and rivers toxic to fish. After initiatives to clean up the air in Europe and North America, Asia is now the most polluted continent, with winter smogs in northern India amongst the worst. A surprising amount of air pollution ends up in the Arctic. Pesticides, carried by the wind, precipitate in the cold air, sometimes reaching toxic levels in birds, whales and polar bears. The global atmosphere is also filling with carbon dioxide. Annual emissions now exceed one tonne for every inhabitant on the Earth, and atmospheric concentrations are enough to cause significant global warming.

Water pollution
Emissions of organic water pollutants

Rank	Country	Grams of pollutant released per worker per day
1	Senegal	360
2	The Bahamas	320
3	Panama	320
4	Mozambique	310
5	St Vincent and the Grenadines	300
6	Malawi	290
7	Armenia	280
8	Guatemala	280
9	Uruguay	280
10	Bermuda	270
11	Ecuador	270
12	Bolivia	250
13	Botswana	250
14	Kenya	250
15	Tanzania	250
16	Chile	240
17	Argentina	230
18	Cyprus	230
19	Ethiopia	230
20	Burkina	220

Air quality in selected Asian cities, 2004

Concentration of PM_{10} particulates (micrograms per m³)

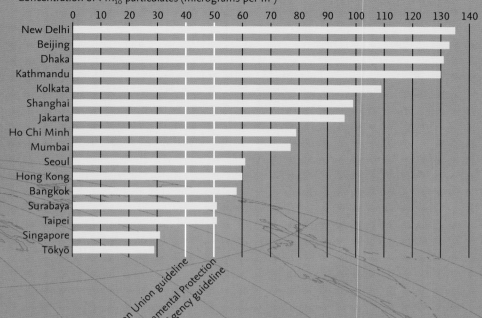

New Delhi
Beijing
Dhaka
Kathmandu
Kolkata
Shanghai
Jakarta
Ho Chi Minh
Mumbai
Seoul
Hong Kong
Bangkok
Surabaya
Taipei
Singapore
Tōkyō

European Union guideline
US Environmental Protection Agency guideline

Many major rivers are heavily contaminated with sewage, industrial chemicals and fertilizer runoff from farms. Many become lifeless, and 'dead zones' are forming in the oceans at the mouths of polluted rivers such as the Mississippi. Meanwhile, soils are contaminated by landfill, factories and air pollution, and irrigated farmland in countries such as Pakistan and Uzbekistan is made infertile by salt contamination. Industrial and transportation accidents can cause major local disasters. The Exxon Valdez oil tanker spill in Alaska in 1989 did long-term damage to fragile Arctic ecosystems. The gas leak from the Bhopal chemicals plant in India in 1984 killed some 4000 people, and fallout from the Chernobyl nuclear fire in Ukraine in 1986 may eventually be responsible for more than 10 000 deaths.

Major pollution incidents 1967–2005

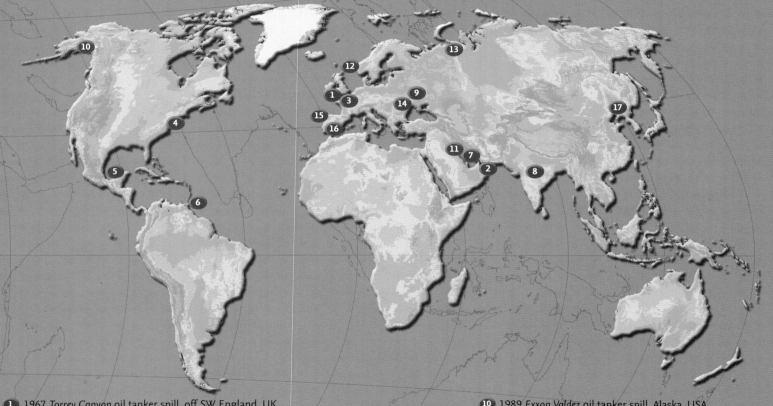

1. 1967 *Torrey Canyon* oil tanker spill, off SW England, UK
2. 1972 *Sea Star* oil tanker spill, Gulf of Oman
3. 1978 *Amoco Cadiz* oil tanker spill, Brittany, France
4. 1979 Radiation leak, Three Mile Island nuclear power station, Pennsylvania, USA
5. 1979 Ixtoc oil well blowout, Gulf of Mexico
6. 1979 Collision of *Atlantic Empress* and *Aegean Captain* oil tankers, off Trinidad and Tobago
7. 1983 Nowruz oil field blowout, The Gulf
8. 1984 Leak of toxic gas, Bhopal, India
9. 1986 Explosion and radiation leak at nuclear power station, Chernobyl, Ukraine
10. 1989 *Exxon Valdez* oil tanker spill, Alaska, USA
11. 1991 Burning oil fields during Gulf War, Kuwait
12. 1993 *Braer* oil tanker spill, Shetland Islands, UK
13. 1994 Ruptured oil pipeline, Usinsk, Russian Federation
14. 2000 Cyanide spill, Baia Mare, Romania
15. 2002 *Prestige* oil tanker spill, off NW Spain
16. 2004 Major forest fires, Spain and Portugal
17. 2005 Benzene spill, Jilin, China

Pristine spruce forest and the effects of acid rain

Acid rain was first identified as an international environmental problem in the 1980s and has affected many parts of the world. Particles released into the atmosphere by burning fossil fuels, by industrial sites and from vehicle exhausts react with sunlight to form harmful chemicals which can be carried long distances by the wind. When these condense, acidic rainwater is formed which in turn can cause severe damage to forests and other vegetation.

A term originally coined in 1905, 'Smog' is a combination of smoke and fog. It relates to instances of severe air pollution where chemicals and harmful particles remain in the air for sustained periods. It causes poor visibility and presents serious health risks. Mexico City is one of the most polluted cities in the world and smog alerts close factories and restrict vehicle use several times a year.

The industrialized nations of the world generate over 2 billion tonnes of waste each year. It is estimated that for each tonne of a finished product, 10 tonnes of waste are created. Such waste can be by-products of industry and commercial activities, domestic rubbish, waste material from mining operations and construction sites and from the generation of nuclear power. In many cases, disposal is not controlled.

causing damage to the environment and also, particularly with toxic waste, presenting great health risks. Even when the waste itself is not

Big

Thaw
Warming world

polar ice – *permanent and seasonal ice sheets, ice caps and sea ice in the polar regions
of the Arctic and Antarctica*

shrinking glaciers – *masses of ice flowing slowly down valleys, whose extent and rate
of movement is affected by climate change*

rising sea level – *average height of the level of the surface of the world's oceans and seas,
likely to rise as a result of global warming*

Antarctica

Area	sq km	sq miles
Total land area *(excluding ice shelves)*	12 093 000	4 667 898
Ice shelves	1 559 000	601 774
Exposed rock	49 000	18 914

Heights	m	ft
Lowest bedrock elevation *(Bently Subglacial Trench)*	-2 496	-8 189
Maximum ice thickness *(Astrolabe Subglacial Basin)*	4 776	15 669
Mean ice thickness *(including ice shelves)*	1 859	6 099

Volume	cubic km	cubic miles
Ice sheet including ice shelves	25 400 000	6 094 000

Climate	°C	°F
Lowest screen temperature *(Vostok Station, 21st July 1983)*	-89.2	-128.6
Coldest Place – annual mean *(Plateau Station)*	-56.6	-69.9
Mean annual temperature at South Pole *(Amundsen-Scott Base)*	-49.5	-57.1
Lowest recorded temperature at South Pole *(Amundsen-Scott Base)*	-66.8	-88.2
Highest recorded temperature at South Pole *(Amundsen-Scott Base)*	-24	-11.2

The northern and southern polar ice caps are very different. Although the Arctic includes land areas such as Alaska, Greenland and Siberia, the North Pole itself sits in the middle of the Arctic Ocean, covered only by drifting sea ice a few metres thick. At the South Pole, near the centre of the Antarctic continent, the presence of a huge landmass has allowed a colossal ice sheet, up to 4 km (2.5 miles) thick, to build up.

The Antarctic supports barely any plant life, whereas the Arctic is host to massive boreal forests and – further north still – a wide extent of treeless tundra. The Arctic is also home to nearly four million people, whereas the southern continent is host only to scientific research stations and a few wandering explorers. Both support abundant animal life, with polar bears in the Arctic and penguins in the Antarctic being the most iconic species.

Polar ice

Combined global land and sea surface temperatures 1856–2004

Relative to 1961–1990 average. The orange line is a smoothing of the annual values.

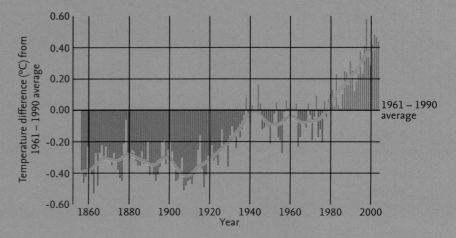

Global warming is exaggerated at the poles, and in the Arctic temperatures have risen at twice the global average rate over the last few decades. The main reason is the 'ice-albedo' effect, where melting snow and ice reveal darker ocean and land surfaces underneath, absorbing more of the sun's energy and causing more warming in a self-reinforcing cycle.

The Antarctic is insulated from wider global changes because it is surrounded by circumpolar ocean currents and winds, maintaining very low temperatures in the interior. Although climate records are sparse, Antarctica's surface is not thought to have warmed significantly over the last thirty years. The exception is the Antarctic Peninsula, which extends north towards South America and is therefore more exposed to the changing climate. The western side of the Peninsula has records going back fifty years, which show an extraordinary 3C° (5.4F°) rise in average temperatures.

The Arctic

Precipitation	mm	inches
Average precipitation (*mainly snow*) **in the Arctic Basin – rain equivalent**	130	5.1
Average precipitation (*mainly snow*) **in the Arctic coastal areas – rain equivalent**	260	10.2

Ice thickness	m	ft
Average sea ice thickness	2	6.6
Maximum ice thickness (*Greenland ice sheet*)	3 400	11 155

Sea ice extent	sq km	sq miles
Minimum sea ice extent in summer	5 000 000	1 930 000
Maximum sea ice extent in winter	16 000 000	6 180 000

Temperature	°C	°F
Summer temperature at North Pole	near 0	32
Winter temperature at North Pole	-30	-22
Lowest temperature recorded in the Arctic (*Verkhoyansk, northeastern Siberia, 1933*)	-68	-90

The changing Arctic

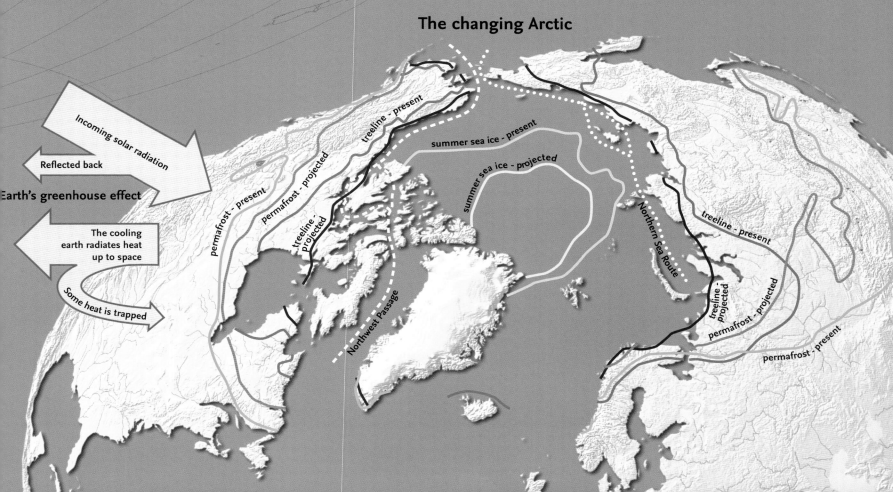

Antarctic profile

Cross-section of West Antarctica from the Ronne Ice Shelf to the Ross Ice Shelf

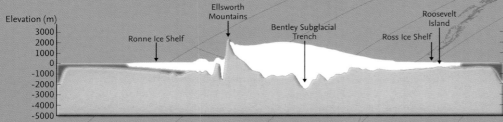

The Larsen A ice shelf in Antarctica collapsed in 1995, with Larsen B following in February–March 2002. The British Antarctic Survey ship *RRS James Clark Ross* navigated her way through an armada of icebergs to collect samples. 'Hard to believe that 500 billion tonnes of ice sheet has disintegrated in less than a month' said BAS glaciologist Dr David Vaughan.

The West Antarctic Ice Sheet has long been a concern to scientists because much of it is grounded below sea level, prompting fears that it could collapse rapidly, adding 5 m (16.5 feet) to global sea levels thereby submerging major cities around the world. Pine Island Glacier, the largest in West Antarctica, has speeded up by 25 per cent since 1970 and is now thinning rapidly.

If all of Greenland melted, it would add 7 m (23 feet) to global sea levels, although it is thought that this would take several centuries. Greenland's major glaciers are already thinning and speeding up, however, suggesting that this meltdown is beginning to happen. Moreover, the surface melt area on the top of the ice cap is also increasing, contributing to rivers and meltwater lakes during the summer months.

The extent of summer sea ice cover in the Arctic Ocean has declined by 15–20 per cent over the past 30 years. September 2005 saw the biggest melt ever recorded. Climate models project a total loss of sea ice later this century, spelling disaster for ice-dwelling animals such as polar bears. In summer 2005 several polar bears were reported to have drowned off Alaska, after sea ice retreated too far offshore for them to swim back.

Snow cover extent has declined by 10 per cent in the last 30 years, and more rainfall is increasing river flows. There is less ice on rivers and lakes, reducing the ice season by three weeks in some areas. Permafrost is melting over wide areas, destabilizing forests and buildings. Indigenous Inuit people in Canada are reporting thinning ice and unpredictable weather, as well as the appearance of previously-unseen birds and fish from the south.

Arctic Ocean profile

Cross-section of the Arctic Ocean from northwest Canada to northwest Russian Federation

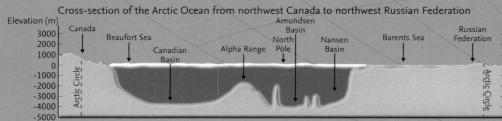

Arctic sea ice concentration   September 1980

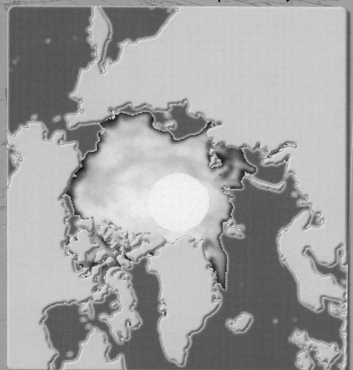

Arctic sea ice concentration   September 2005

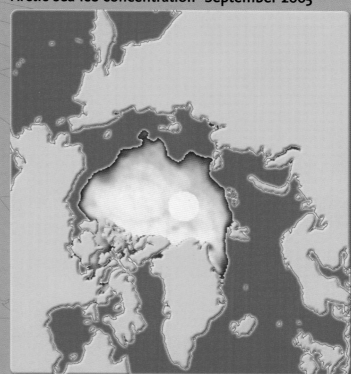

Antarctic sea ice concentration   February 1980

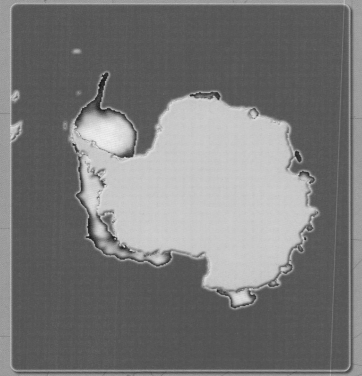

Antarctic sea ice concentration   February 2005

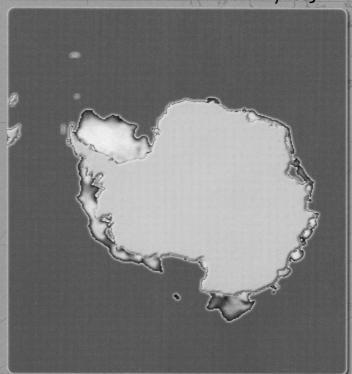

per cent

0 5 10 15 20 25 30 35 40 45 50 55 60 65 70 75 80 85 90 95 100

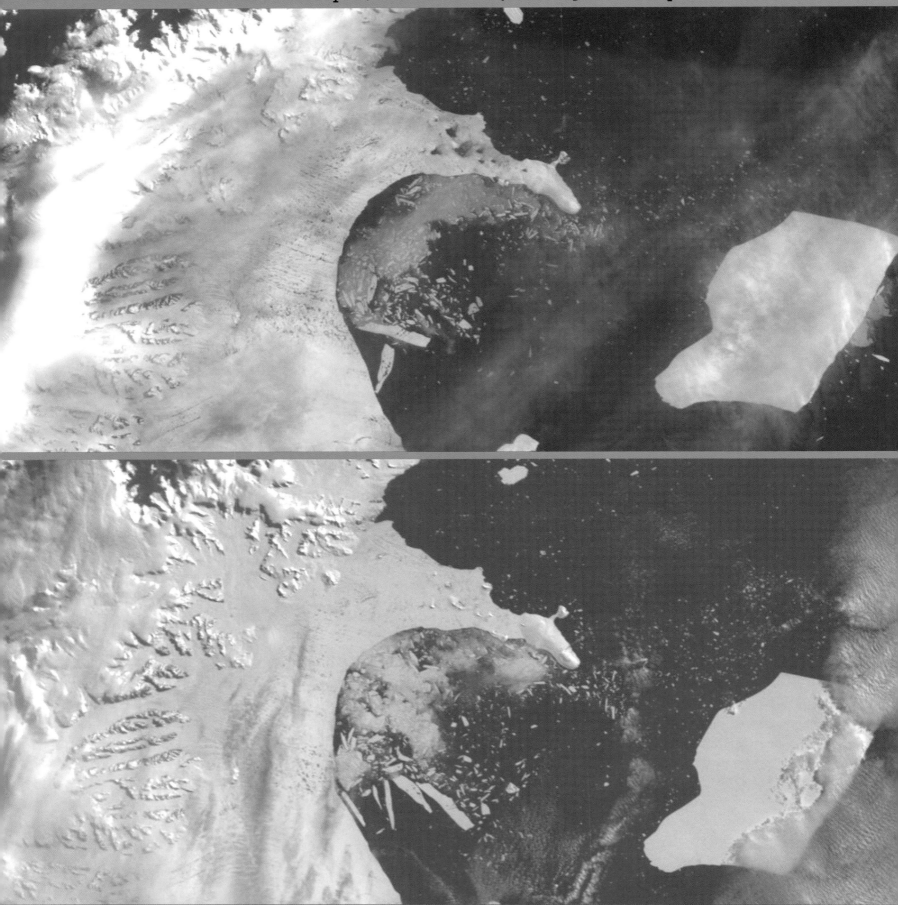

An ice shelf typically advances for several decades until it becomes unstable and icebergs break off, or calve, from the front of the shelf. This advance and retreat is normal and maintains the ice volume. Scientists had predicted that a retreat was due to happen to the Larsen B ice shelf. In the Antarctic summer of 2002, rather than calving, it completely disintegrated over a 35-day period. It released nearly 3000 sq km

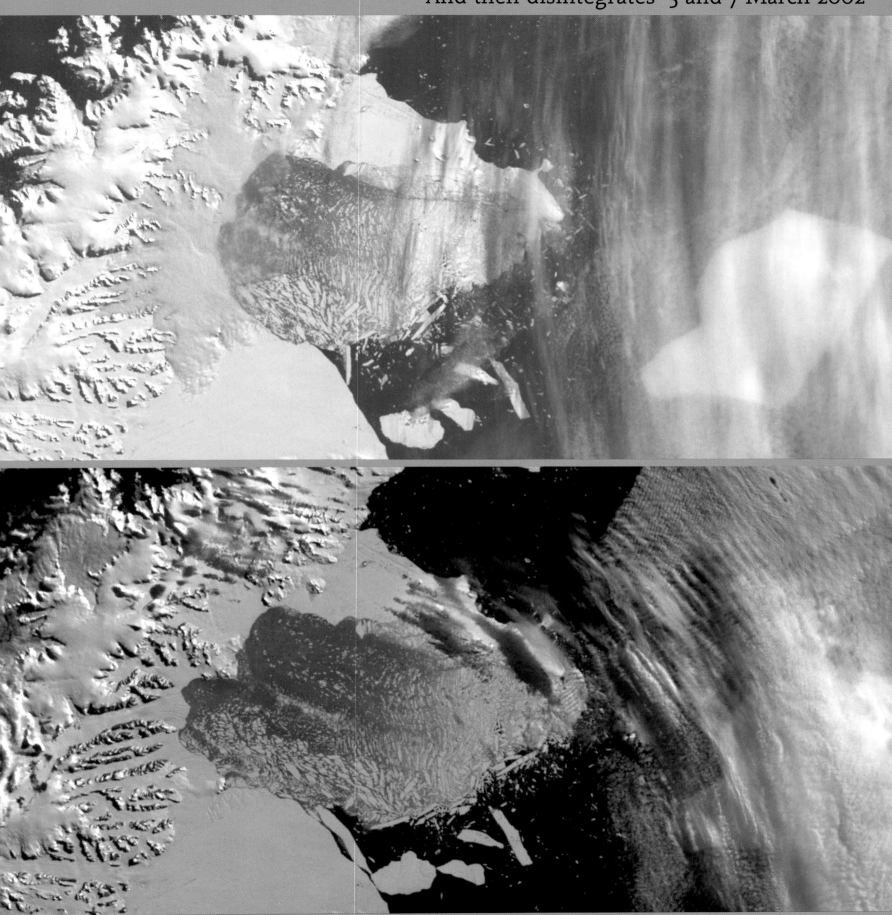

(1158 sq miles) of ice, equivalent to an area larger than Luxembourg. This is the largest ice shelf retreat in the Antarctic Peninsula in the last 30 years and is attributed to significant warming of the local climate since the late 1940s.

The ice sheet covering Greenland has an area of 1 833 900 sq km (708 069 sq miles) and an average thickness of 2.3 km (1.4 miles). It is the second-largest concentration of frozen freshwater on Earth and if it were to melt completely the global sea level would rise by up

Expanding inland 14 June 2005

The rapidly expanding melt zone (highlighted) on the western edge of the ice sheet can be seen where the water has darkened the ice to blue-grey. June is in the melting season which reaches its peak in late August or early September. Fortunately, the ice in the interior of the island is still accummulating.

Frozen in winter ice, Shishmaref, Alaska, USA 2003

This is a winter view of Shishmaref, a village on Sarichef Island in the Chukchi Sea, just north of the Bering Strait and five miles from mainland Alaska. At this time of the year the sea ice gives protection to the shoreline. In the past it also protected the shore from storm surges for much of the year but now in the summer there is much less ice and the coastline is suffering as a result.

Effects of melting permafrost, Shishmaref, Alaska, USA September and October 2005

The permafrost which the village is built on is also melting. This makes the shore much more vulnerable to erosion. Recent erosion rates average around 3.3 m (10 feet) per year and buildings are being lost. The community is now faced with moving the village's location or building sea walls to give themselves some of the protection once afforded by the sea ice.

Permafrost melt has several consequences for arctic lands. In towns, buildings once stable on frozen footings are now slipping towards one another as their foundations sink into soft ground. Similarly, in the forests the softened ground affects trees which slide 'drunkenly' into one another or collapse completely due to lack of support.

The effect of melting ice on polar bears is significant. They are forced to spend longer on land, the ice floes they Arely on to move Around are reduced and they can drown or become stranded and starve. Antarctic penguins are also suffering a reduction in numbers. Chinstrap penguins are fortunate because they prefer open water and have managed to increase their numbers.

Global distribution of glaciers and ice sheets

Glaciers numbered are those shown in the graph on the opposite page

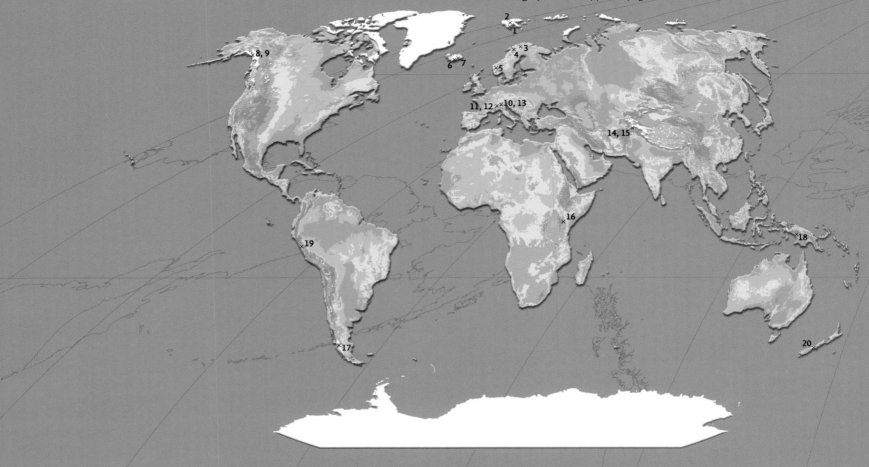

Cumulative changes in global glacier thickness and glacier contributions to sea level rise.

Glaciers form wherever it is cold enough for snowfall to accumulate into thick layers of ice over many years rather than melting away during the summer. Areas cold enough to support permanent ice may be close to sea level at high latitude regions such as Patagonia, but only exist at high altitudes of 5000 m (16 404 feet) or more in hot tropical areas such as Peru.

Glaciers expand and contract according to the balance between temperature and precipitation. Like a bank account, withdrawals from melting must be balanced by deposits from new snowfall or the glacier will shrink. Glaciers also shrink if the melt area on their surface increases because rising temperatures push the snowline higher up the mountainside. This is the main way in which global warming is affecting glaciers.

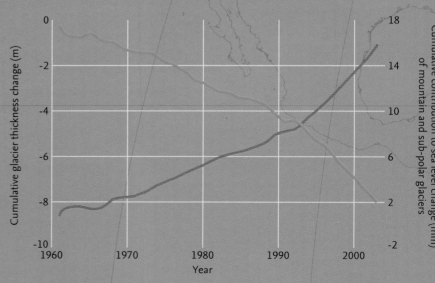

Cumulative contribution to sea level change of mountain and subpolar glaciers

Cumulative glacier thickness change

Shrinking glaciers

Global glacier retreat

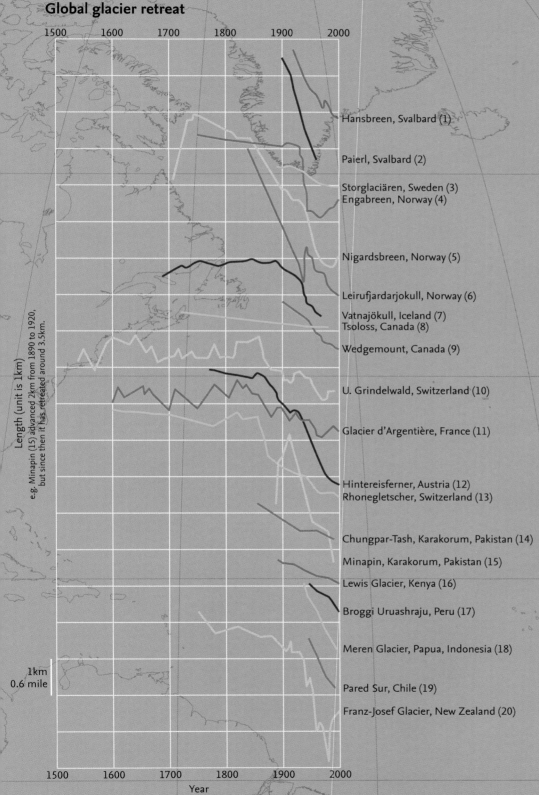

e.g. Minapin (15) advanced 2km from 1890 to 1920, but since then it has retreated around 3.5km.

Length (unit is 1km)

Hansbreen, Svalbard (1)

Paierl, Svalbard (2)

Storglaciären, Sweden (3)
Engabreen, Norway (4)

Nigardsbreen, Norway (5)

Leirufjardarjokull, Norway (6)

Vatnajökull, Iceland (7)
Tsoloss, Canada (8)

Wedgemount, Canada (9)

U. Grindelwald, Switzerland (10)

Glacier d'Argentière, France (11)

Hintereisferner, Austria (12)
Rhonegletscher, Switzerland (13)

Chungpar-Tash, Karakorum, Pakistan (14)

Minapin, Karakorum, Pakistan (15)

Lewis Glacier, Kenya (16)

Broggi Uruashraju, Peru (17)

Meren Glacier, Papua, Indonesia (18)

Pared Sur, Chile (19)

Franz-Josef Glacier, New Zealand (20)

1km
0.6 mile

Year

According to the World Glacier Monitoring Service, eighty-three out of eighty-eight glaciers saw a negative 'mass balance' for the year 2003–4, with glaciers shrinking in countries as far apart as Argentina and Norway. The extra meltwater generated is already making a measurable contribution to sea level rise. Africa's highest mountain, Kilimanjaro, has already lost 80 per cent of its ice over the last century, and if melting continues at the current rate, all of its glaciers will have disappeared by 2015.

In Peru, the Andean glaciers have lost around a third of their area in recent decades. The loss of glacial ice is of particular concern to the Peruvian people because their dry season freshwater supplies depend upon glacial run off from the mountains. One Peruvian glacier, the Qori Kalis, is now retreating at 155 m (509 feet) every year – 32 times faster than the rate of retreat in the 1960s and 1970s (see page 170). Glaciologist Lonnie Thompson, who has drilled ice cores in the ice cap which feeds Qori Kalis, calls the rate of retreat 'incredible'. He compares tropical glaciers to the use of canaries in a coal mine: 'They're an indicator of massive changes taking place'.

Upsala Glacier, Argentina 1928 and 2004

The Upsala Glacier is one of the largest Patagonian glaciers. It is a 'calving' glacier which means that when the front of the glacier is in contact with water it loses a significant part of its mass due to large pieces of ice falling off. All parts of this glacier retreated after 1978. However, not all Patagonian glaciers have retreated and the Moreno Glacier, also in Argentina, has even advanced.

Part of Upsala continued to retreat until 1999 but since then it has shown seasonal advances and retreats over a distance of about 400 m (1312 feet). A comparison of recent glacier velocity to 30 years ago suggests that the retreat was due to a combination of release of backstress pressure along with the glacier thinning due to local temperature increases.

Qori Kalis Glacier, Peru 1978 and 2002

Qori Kalis glacier is the largest outlet from the Quelccaya Ice Cap in southeast Peru. The ice cap has shrunk by approximately 20 per cent since 1963 as the snowline has retreated due to increasing temperatures. Local water supplies and lakes are drying up because of the rising temperatures and lower levels of precipitation. However, in meltwater season there are floods and landslides. The glacier itself is retreating at an alarming rate. Between 1998 and 2001 it's retreat averaged 155 m (509 feet) per year, 32 times faster than between 1963 and 1978.

Glacier Toboggan is in the Cugach Mountains south of Anchorage, and drains into Harriman Fjord. It was named in 1899 by members of the Harriman Alaska Expedition and like most Alaskan glaciers it has shrunk significantly since the early twentieth century as a result of climate change. Loss of glaciers is contributing to an increase in the frequency of earthquakes in Alaska as the loss of their weight allows tectonic plates to move more easily.

Alaska's Glacier Bay has been a National Monument since 1925 and a National Park since 1980. One of the glaciers to be found there is the Muir Glacier which is named after John Muir, the naturalist and explorer who first viewed it on his 1879 expedition. The southern Alaskan glaciers are sensitive to climate change and many have shrunk or disappeared over the last 100 years.

The Muir Glacier is now only a shadow of its former glory, and has nearly retreated up out of the ocean. Many of the large tidewater glaciers that John Muir first saw in 1879 have become small glaciers terminating on land. As the glaciers recede, plants and animals are recolonizing the area. The retreat seems to be caused by higher temperatures and changes in precipitation.

Kilimanjaro, Tanzania A snow covered peak in 1974

Kilimanjaro literally means 'the mountain that glitters'. It is one of Africa's most stunning landscapes, and after his visit in 1933 Ernest Hemingway wrote the 1938 short story *The Snows of Kilimanjaro*. The book and the 1952 film which followed have encouraged many thousands of visitors to Tanzania.

During the last few decades, the permanent snow and ice on the summit of Kilimanjaro has almost completely vanished. This loss is mainly due to locally increasing average annual temperatures. The glaciers could be completely gone from Kilimanjaro in the next 10 to 15 years. The ice cap formed more than 11 000 years ago and 80 per cent of the ice fields have been lost in the last century.

Gangotri Glacier, India 2001

This false-colour satellite image shows the Gangotri Glacier in northern India. At its current length of 30.2 km (18.7 miles) it is one of the longest in the Himalaya. The glacier is the source of the Bhagirathi river, an important tributary of the Ganges river. It is also a place of traditional Hindu pilgrimage. The Gangotri Glacier has been receding since 1780, and the retreat quickened after 1971. Over the last 25 years

he glacier has retreated more than 850 m (2789 feet), with a recession of 76 m (249 feet) from 1996 to 1999 alone. This is a concern as the glacial channel which feeds the river has changed course and the volume of water is shrinking rapidly, mainly due to reduced winter precipitation. Local deforestation around the glacier is also adding to the problem and it is feared that parts of the glacier may disintegrate.

The movement of the Pasterze Glacier is one of many well-documented in Europe. The measurements taken show a retreat which actually started as far back as 1856. This 1875 image shows how extensive the glacier was at that date, but a more recent combination of higher summer temperatures and lower winter snowfall has increased the rate of retreat.

...no longer visible in 2004

By 2004 the glacier had retreated so much that it could no longer be seen from the same viewpoint. Recent measurements of the glacier terminus show that it has shrunk every year since the winter of heavy snow in 1965–1966. Between 1986 and 1990 the average annual retreat was 15 m (49 feet). This then increased so that by 2000 the average distance lost each year was 19 m (62 feet).

Franz Josef Glacier, New Zealand retreats...

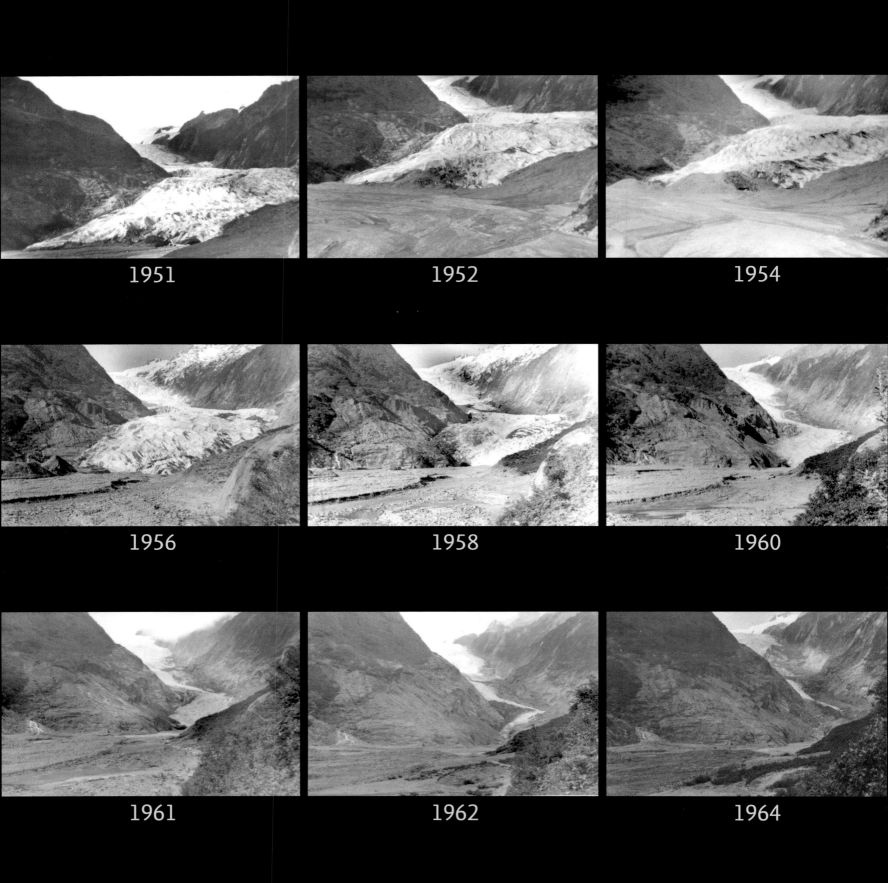

Franz Josef Glacier has shown a well-documented retreat over a number of years. The previously gradual retreat of the glacier was significantly accelerated by a number of warm winters and in this series of images it can be seen vanishing off into the distance in the middle of a period of rapid retreat which lasted over forty years.

In the 1980s the snow conditions in this area changed. This reversed the retreat and the glacier is now advancing again, as can be seen here in 2001. Access, which had been made difficult due to rock falls and dangerous river-crossings, has improved and there is again a huge demand by visitors for guided trips in this area.

Trends in mean sea level 1993–2005

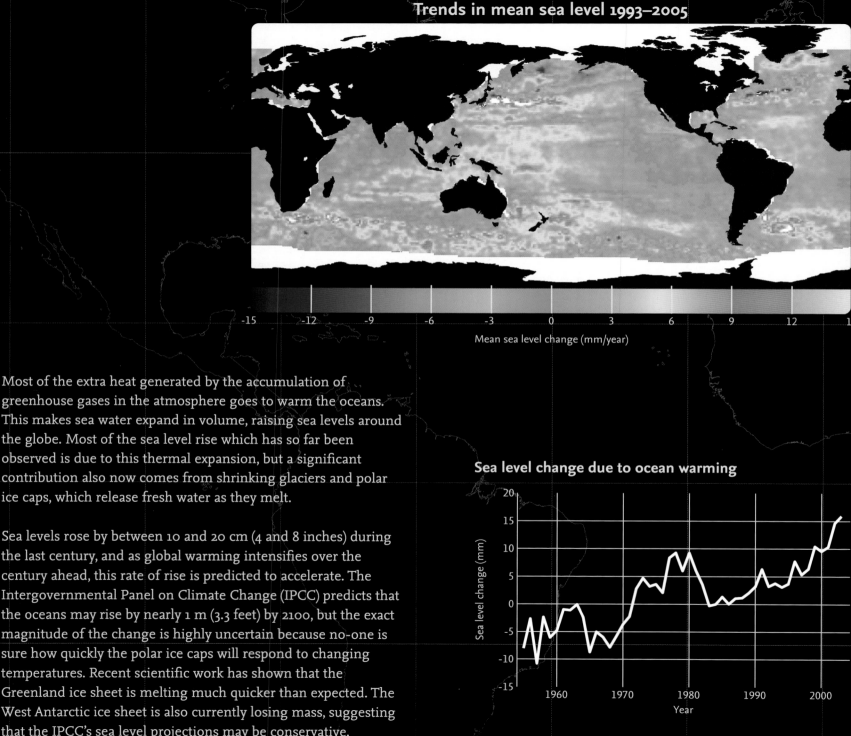

Mean sea level change (mm/year)

Most of the extra heat generated by the accumulation of
greenhouse gases in the atmosphere goes to warm the oceans.
This makes sea water expand in volume, raising sea levels around
the globe. Most of the sea level rise which has so far been
observed is due to this thermal expansion, but a significant
contribution also now comes from shrinking glaciers and polar
ice caps, which release fresh water as they melt.

Sea levels rose by between 10 and 20 cm (4 and 8 inches) during
the last century, and as global warming intensifies over the
century ahead, this rate of rise is predicted to accelerate. The
Intergovernmental Panel on Climate Change (IPCC) predicts that
the oceans may rise by nearly 1 m (3.3 feet) by 2100, but the exact
magnitude of the change is highly uncertain because no-one is
sure how quickly the polar ice caps will respond to changing
temperatures. Recent scientific work has shown that the
Greenland ice sheet is melting much quicker than expected. The
West Antarctic ice sheet is also currently losing mass, suggesting
that the IPCC's sea level projections may be conservative.

Sea level change due to ocean warming

Changing sea levels are important because one-fifth of humanity lives within 30 km (18.6 miles) of the ocean. Major cities at risk from rising waters include Shanghai, New York, Mumbai and Tōkyō. Meanwhile, low-lying atolls such as in the Pacific island nations of Tuvalu and Kiribati are particularly vulnerable because their entire land area could be swamped in just a few decades. Already people in Tuvalu report a higher incidence of flooding during high tides than was observed in previous years, and plans are being made to evacuate the population to New Zealand. According to Paani Laupepa, of Tuvalu's environment ministry: 'This is one of the biggest threats that has ever faced our nation, and I think the entire world.' He asks: 'How do you put a price on an entire nation being relocated?'

Lowest Pacific islands

	Maximum height above sea level	Land area sq km	sq miles	Population
Kingman Reef	1 m (3 ft)	1	0.4	0
Palmyra Atoll	2 m (7 ft)	12	5	0
Ashmore and Cartier Islands	3 m (10 ft)	5	2	0
Howland Island	3 m (10 ft)	2	1	0
Johnston Atoll	5 m (16 ft)	3	1	0
Tokelau	5 m (16 ft)	10	4	1 000
Tuvalu	5 m (16 ft)	25	10	10 000
Coral Sea Islands Territory	6 m (20 ft)	22	8	0
Wake Island	6 m (20 ft)	7	3	0
Jarvis Island	7 m (23 ft)	5	2	0

Threat of rising sea level

Major cities
Coastal areas at greatest risk
Islands and archipelagos
Areas of low-lying islands

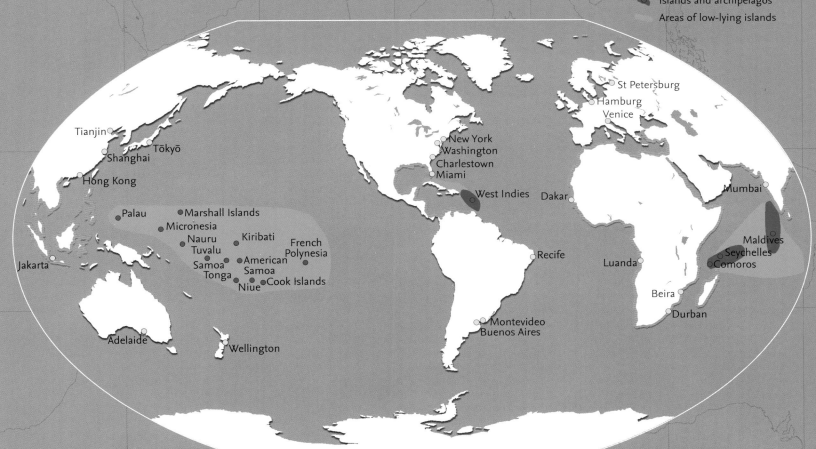

Low tide in Tuvalu

Tuvalu is an island nation in the Pacific Ocean. Its inhabitants live mainly on coral atolls which are very low-lying. Many of its islands are only a few metres above sea level at their highest point. Over recent years the islanders have seen many changes in their fortunes, but the most important issue now to most is that as the tide comes in, their property is at risk of flooding.

Warming of the oceans has raised sea levels in parts of the Pacific and certain low-lying island groups such as Tuvalu are now very vulnerable. They face the prospect of losing their national identity as more and more inhabitants relocate to countries such as New Zealand

The Maldives, Indian Ocean...

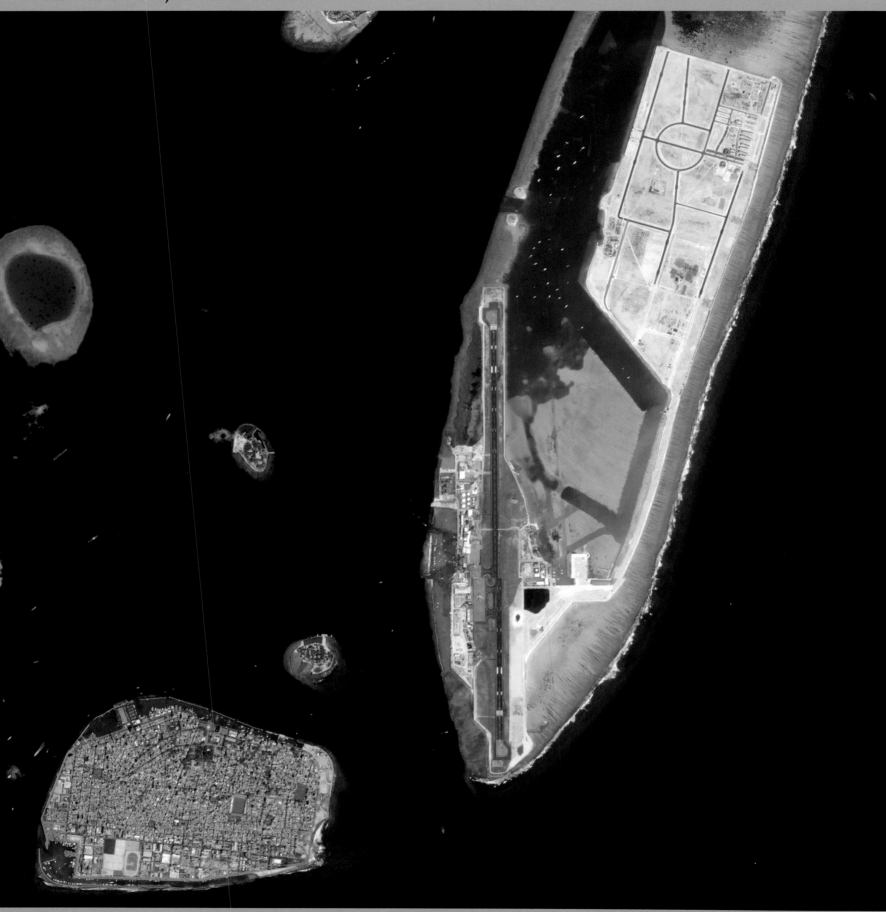

In the Maldives, in the Indian Ocean, storms do not raise the water level by more than around 300 mm (11.8 inches). Accordingly, development has been concentrated on land only 40 cm (15.7 inches) or more above sea level. This means that they are very vulnerable to sea level rise. Most of the population lives within 2 m (6.5 feet) of sea level and almost all within 4 m (13.1 feet).

Male, the capital of the Maldives, is approximately 2 m (6.5 feet) above the sea, but its reclaimed land is lower. After storms in 1987 and 1988 flooded the reclaimed areas, a series of breakwaters on the outer coast were built to protect the town from damaging storm waves, but they will not prevent flooding from a sustained rise in sea level.

Hatteras Airfield, North Carolina, USA 17 July 1996 and 8 August 1999

The east coast of the USA is affected by rising sea levels, damage from hurricanes, and other storms, all of which combine to change the coastline and to make parts of it very vulnerable to the sea at high tide. The lower post-hurricane image shows how quickly this can happen in an area where barrier islands are actively moving westwards toward the mainland.

The famous Cape Hatteras lighthouse is the tallest brick lighthouse in the USA at 63 m (208 feet). It was built 488m (1600 feet) from the sea in 1870. Because the barrier islands are moving westwards it could not be protected from the sea indefinitely so in 1999 it was moved a similar distance inland.

Parched

Earth

Deserts and drought

dvancing deserts – *the encroachment of desert conditions into settlements or agricultural areas as a result of climate change or bad farming practices*

rought and fire – *prolonged period of dry weather leading to water shortages, loss of crops and an increased risk of fire*

hrinking lakes, drying rivers – *reduction in the size of lakes and river flow rates due to climate change or the extraction of water for agriculture or industry*

Each spring, billowing clouds of dust sweep down from the drylands in the north of China and envelop Beijing. Homes, streets and cars are left coated with brown dirt, and the skies turn an opaque orange colour. Sand dunes are now a mere 70 km (43.5 miles) from Beijing itself, and huge areas of Inner Mongolia, Gansu and Xinjiang provinces are being engulfed by expanding desert.

But the Chinese are fighting back. Efforts have been made to stop the overgrazing by sheep and goats which triggers desertification by removing grass and bushes, which lets strong winds erode the sandy soil. Ploughing has also been banned in areas where it allows topsoil to blow away. The Chinese government is also planting a 'green wall' of trees – almost as long as the famed Great Wall of China – to try to stop the advancing sands. It claims that anti-desertification measures have slowed the rate of desertification from 3400 sq km (1313 sq miles) to 1200 sq km (463 sq miles) per year.

North and South America 27%

In Mexico 47% of the land is affected by desertification causing vast population movements.

Advancing deserts

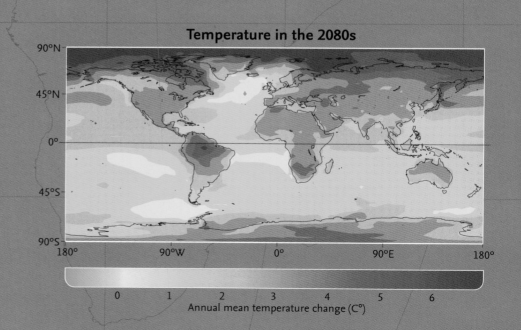

Temperature in the 2080s

Annual mean temperature change (C°)

Land at risk of desertification

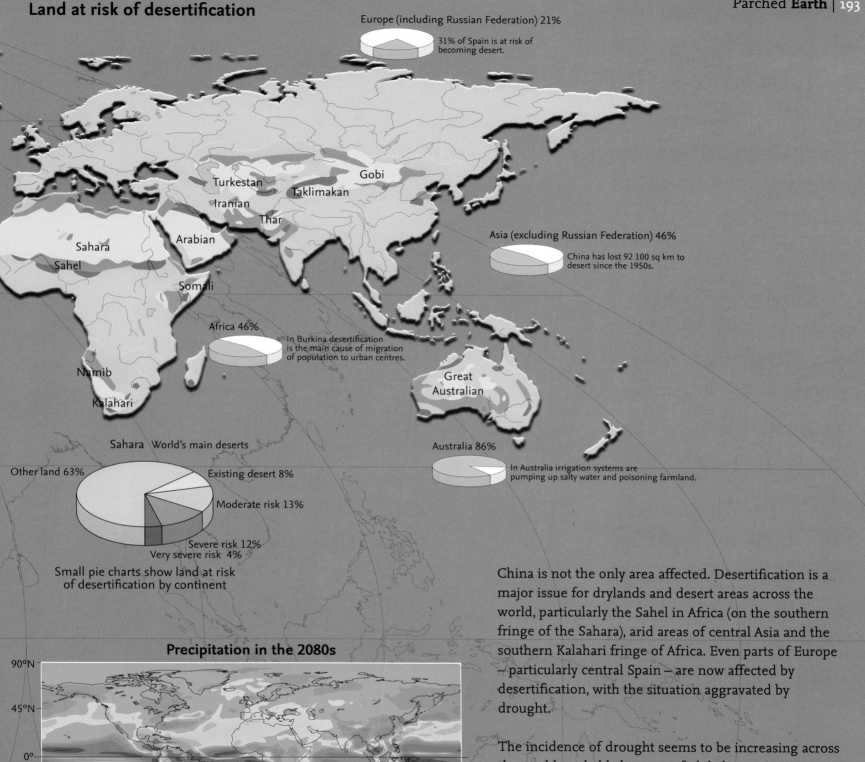

Europe (including Russian Federation) 21%

31% of Spain is at risk of becoming desert.

Turkestan

Gobi

Iranian

Taklimakan

Thar

Sahara

Arabian

Sahel

Somali

Asia (excluding Russian Federation) 46%

China has lost 92 100 sq km to desert since the 1950s.

Africa 46%

In Burkina desertification is the main cause of migration of population to urban centres.

Namib

Great Australian

Kalahari

Sahara   World's main deserts

Other land 63%

Existing desert 8%

Moderate risk 13%

Severe risk 12%

Very severe risk 4%

Small pie charts show land at risk of desertification by continent

Australia 86%

In Australia irrigation systems are pumping up salty water and poisoning farmland.

Precipitation in the 2080s

90°N

45°N

0°

45°S

90°S

180°   90°W   0°   90°E   180°

-3   -2   -1 -0.5 -0.25 0 0.25 0.5 1   2   3

Average precipitation change (mm per day)

China is not the only area affected. Desertification is a major issue for drylands and desert areas across the world, particularly the Sahel in Africa (on the southern fringe of the Sahara), arid areas of central Asia and the southern Kalahari fringe of Africa. Even parts of Europe – particularly central Spain – are now affected by desertification, with the situation aggravated by drought.

The incidence of drought seems to be increasing across the world, probably because of global warming. Scientific studies have linked droughts in Africa and the United States to rising ocean temperatures which change weather patterns in complex ways. Computer model projections for climate change over the next century suggest that this problem will get dramatically worse – as temperatures rise further, new deserts will spread across southern Africa, central North America and the Mediterranean fringe of Europe.

Many parts of the world, on the delicate fringes between fertile and arid regions, are at risk from advancing deserts. This is shown graphically in this image of the Sahel region of Africa. The Sahel is a buffer zone between the Sahara Desert to the north and the savannah grasslands to the south. Sand dunes are also threatening oases in the desert and various measures are being taken to try to restrict the advance of sand onto

valuable arable and cattle-rearing land. Climate change and man's destruction of natural vegetation are exacerbating the problem and threatening the traditional way of life in many areas.

Egypt and the Nile 19 July 2004

The Nile Valley in Egypt illustrates the fine line between desert and fertile land. Completely surrounded by desert, the Nile is a lifeline for the Egyptian people. Until the Aswan Dam was built the Egyptians relied on the annual flooding of the Nile to irrigate their land but this flooding was not guaranteed and drought and famine were commonplace.

Where water can be found, so too can animals and people. Oases offer a degree of stability to the population. However, they are also a destination for the nomadic cattle farmers who eke out an existence in the scrublands that border the deserts. These images show typical scenes around an Algerian oasis and in the Sahel region of the Lake Chad Basin, Niger.

There are many ways of controlling advancing deserts, such as planting trees and nitrogen-fixing plants, or spraying petroleum over seeded land to prevent moisture loss. Or more simply, as along the Trans Saharan Highway in Mauritania, by erecting fences to slow down the encroachment of the sand dunes.

One of the areas most threatened by advancing desert is the land which lies to the east of the Gobi Desert in China. The village of Langtou Gou lies only 130 km (81 miles) from the capital Beijing and it is being threatened by sand blowing from the Gobi, destroying fields and covering houses.

Drought and fire

Major droughts 1964–2005

Year	Country	Deaths
2005	Burundi	120
2005	Kenya	27
2002	Malawi	500
2001	Pakistan	80
2001	Angola	58
2001	Guatemala	41
2001	India	20
2000	Pakistan	143
2000	Uganda	115
2000	Kenya	85
2000	Somalia	21
1998	Indonesia	212
1998	Papua New Guinea	28
1997	Indonesia	460
1997	Papua New Guinea	60
1988	China	1 400
1987	Somalia	600
1987	Ethiopia	367
1987	India	300
1987	India	110
1987	Mozambique	50
1986	Indonesia	84
1984	Ethiopia	300 000
1984	Sudan	150 000
1984	Mozambique	100 000
1984	Chad	3 000
1984	Indonesia	230
1983	Swaziland	500
1983	Brazil	20
1982	Indonesia	280
1978	Indonesia	63
1974	Ethiopia	200 000
1974	Somalia	19 000
1973	Ethiopia	100 000
1967	India	500 000
1967	Australia	600
1966	India	500 000
1966	Indonesia	8 000
1965	India	500 000
1965	Ethiopia	2 000
1964	Somalia	50

Fire is a natural part of the ecosystem in many parts of the world. In Australia, many species of eucalyptus trees require their seed pods to be scorched by fire before they will open and germinate. Australia's Aboriginal people used 'firestick farming' to clear grasslands and manage their hunting grounds.

Fire does not sit well with heavily-populated urban areas, however, where millions of dollars' worth of real estate – not to mention a family's entire possessions – can go up in flames once a wildfire gets out of control. The worst fires can also be deadly: in 1983 twelve firecrew members were killed near Melbourne in a feared 'crown fire' which jumped between tree tops. In 1994 Sydney was almost encircled by over 800 separate fires, which rained ash onto the central business district and shut out the sun as brown smoke drifted over vast areas.

Fire and drought are closely linked. The worst fires of the modern era struck the forests of Indonesia during the drought sparked by El Niño in 1998. The conflagrations not only blotted out entire countries with smoke and smog, causing respiratory diseases and grounding aircraft, but released millions of tonnes of carbon dioxide from burning underground peat and above-ground vegetation, worsening climate change.

Humans are directly involved in this process. The Indonesian fires were partly caused by farmers clearing land for palm oil plantations. The 2003 wildfires in Portugal were caused by the hottest temperatures to hit Europe for centuries, but were sparked in each case either by carelessness or arson. When the flames finally died down, eighteen people were dead and an area the size of Luxembourg had been devastated.

There is little doubt that global warming will make fires worse. The Mediterranean area can expect the burning season to lengthen by several weeks. Australia and California will also be severely affected, while areas in the far north such as Alaska – where forest fires are normally a rarity – will also see huge areas turned to ashes.

California 1994 2003 2004 · Arizona 2000 · New Mexico 2000 · Florida 1998 · Guatemala 2001 · Roraima 1998 · Brazil 1983 · Argentina 1987

Major wildfires 1967–2005 and droughts 1964–2005

Mongolia
1996

Sakhalin
1998

Heilongjiang
1987

China
1988

...tugal
2003

Spain
1994

Pakistan
2000, 2001

India
1965, 1966, 1967,
1987, 2001

Chad
1984

Sudan
1984

Ethiopia
1965, 1973, 1974
1984, 1987

Somalia
1964, 1964,
1974, 1987, 2000

Myanmar
1997

Uganda
2000

Burundi
2005

Kenya
2000, 2005

Sumatra
1997

Indonesia
1966, 1978, 1982, 1984,
1986, 1997, 1998

Papua
New Guinea
1997, 1998

Angola
2001

Malawi
2002

Indonesia
1994

Mozambique
1984, 1987

Swaziland
1983

Australia
1967

New South Wales
1994, 1994

South Australia
1983

Canberra
2001
2003

Tasmania
1967

▢ Droughts
● Wildfires

Major wildfires 1967–2005

Year	Location	People affected	Area affected sq km	sq miles	Year	Location	People affected	Area affected sq km	sq miles
2004	California, USA	15 000	66	25	1994	California, USA	1 200	10 000	3 861
2003	Canberra, Australia	2 650	30 000	11 583	1994	New South Wales, Australia	26 020	8 000	3 089
2003	Portugal	150 000	4 200	1 622	1994	Spain	15 020	2 700	1 043
2003	California, USA	27 104	1 133	437	1994	Indonesia	3 000 000	1 360	525
2001	Canberra, Australia	4 400	5 500	2 124	1994	New South Wales, Australia	20 141	unknown	
2000	Arizona, USA	1 000	680 000	262 550	1992	Nepal	50 000	unknown	
2000	New Mexico, USA	25 400	194	75	1987	Heilongjiang, China	56 313	25 000	9 653
1998	Roraima, Brazil	12 000	9 254	3 573	1987	Argentina	152 752	unknown	
1998	Sakhalin, Russian Federation	100 683	5 000	1 931	1983	South Australia, Australia	11 000	unknown	
1998	Florida, USA	40 124	1 052	406	1981	Myanmar	48 588	unknown	
1997	Sumatra, Indonesia	32 000	800	309	1967	Tasmania, Australia	3 100	unknown	
1996	Mongolia	5 061	80 000	30 888					

Dry lake bed, Lake Amboseli, Kenya

Nothing conjures up the image of drought more than herds of cattle kicking up dust as they cross a dried up dusty lake bed in search of water and food. This image of Masai farmers driving their cattle across a dusty Lake Amboseli is typical of drought-threatened lands throughout the world.

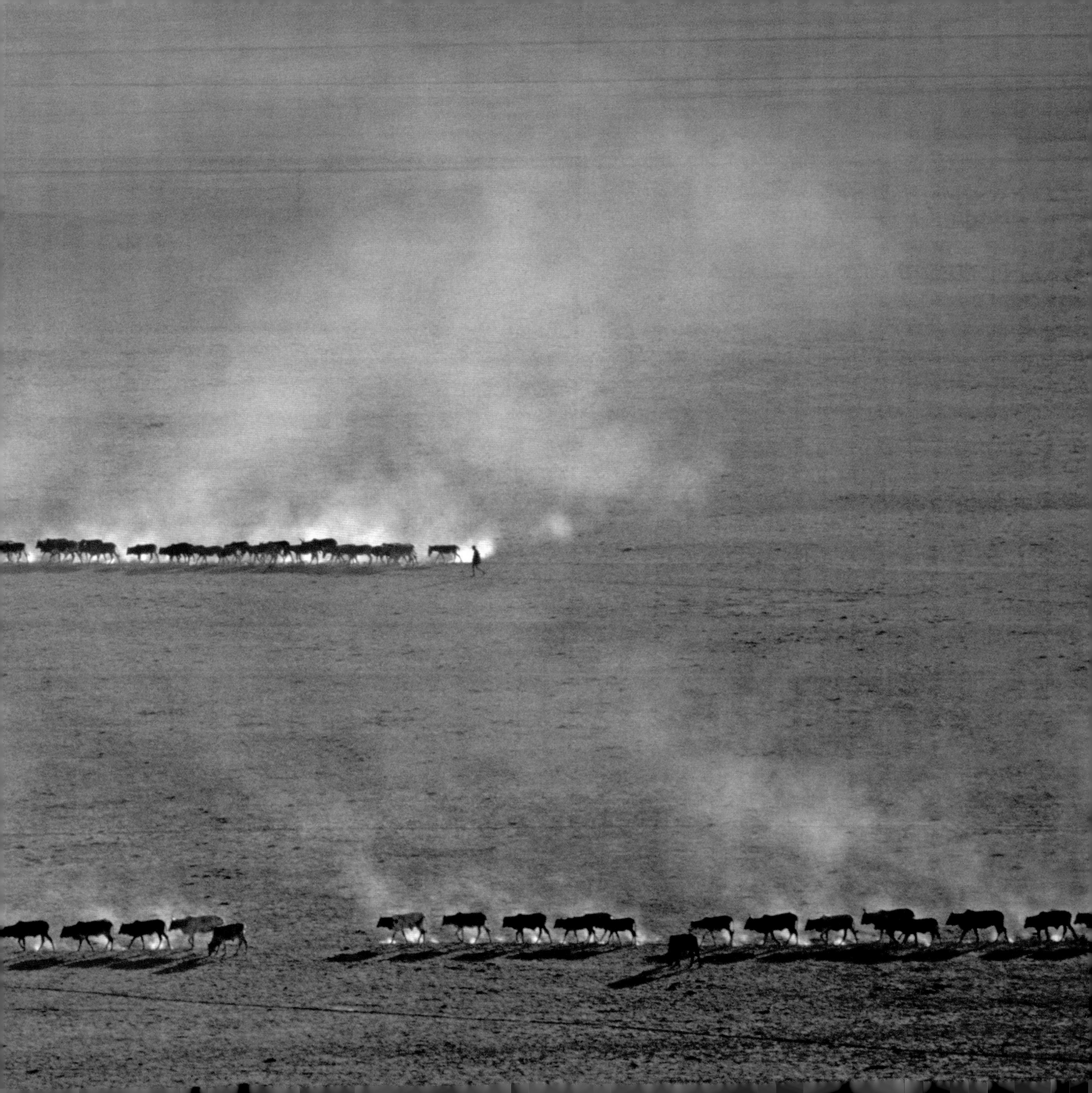

Drought, Australia 2005

January

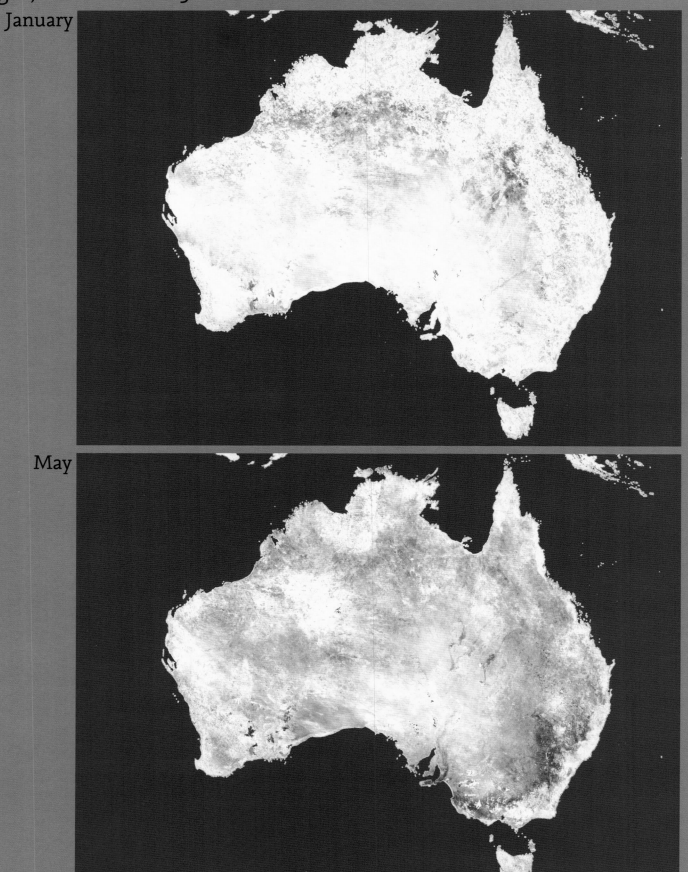

May

These two images have been created from vegetation data from the SPOT satellite. Areas where vegetation is healthy appear green; areas where it is sparse And unhealthy as a result of lack of water appear brown. In summer (January) much of the interior of Australia is brown but by early winter (May), when one would expect more green to appear, the majority of the continent is in the grip of a drought with only the southwestern region showing any healthy vegetation. The darker the brown areas appear, the drier than normal they are.

Drought affects both man and animals. The Australian drought resulted in waterholes drying up, with wild animals forced to seek out any available water. In Burkina, the lowering of the water table necessitates the deepening of wells. In the right-hand image the man's partner is 10 m (33 feet) below the surface digging the well.

Wolf Lake, Yellowstone National Park, USA 1988

The summer of 1988 was one of the driest in Yellowstone National Park's recorded history and a series of wildfires scorched over 4800 sq km (1853 sq miles) of the Park. Over 200 fires started in the greater Yellowstone region, resulting in $3 million damage to property and the deaths of a number of elk, deer, moose, black bear and bison.

Although the aftermath was a scene of devastation the natural process of regeneration began almost immediately. Although the surface of the earth was burnt, much of the plant root systems and seeds remained unharmed and in the following years these plants began to regrow. There is historical evidence that wildfire can play an important role in an area's ecosystem with many plants able to tolerate fire.

Trend in level of Lake Chad 1870–2000

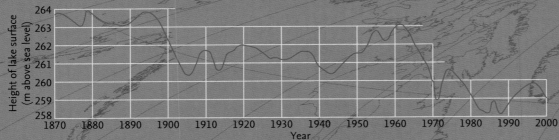

Shrinking lakes, drying rivers

Humans have always based settlements near lakes and rivers. These water bodies provide fresh water, fisheries, transportation and irrigation for crops. But in recent decades, as urban and agricultural demands for water have grown, rivers and lakes across the world have been under assault.

Many of the world's largest rivers now see their flows controlled by people rather than nature. In the United States, the Colorado and Rio Grande river systems have big dams in their catchments which channel water for irrigation and to big desert cities such as Las Vegas. As a result, no water reaches the sea from either river for much of the year. In China, the Yellow River also runs dry most of the time in its lower reaches.

The two rivers which feed into the Aral Sea in Central Asia were dammed in the 1960s to provide water for cotton plantations, devastating the whole area. What was once the fourth-largest inland lake in the world has now retreated into three sections, leaving huge areas of dusty lakebed and old fishing ships stranded on dry land. Only the northern section is thought to be salvageable – the southern Aral Sea will soon dry up for ever.

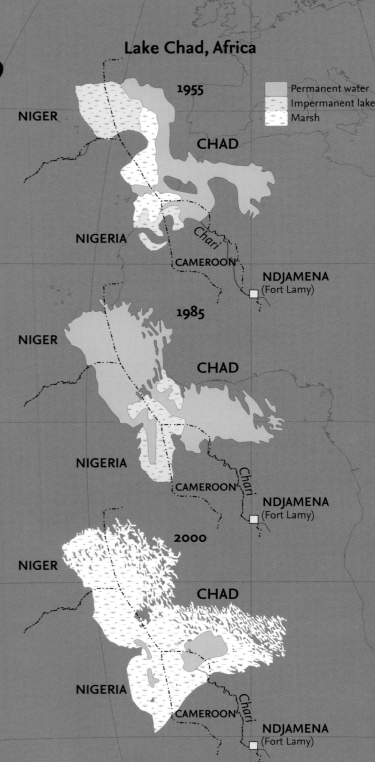

Lake Chad, Africa

Permanent water
Impermanent lake
Marsh

1955
NIGER
CHAD
NIGERIA
CAMEROON
Chari
NDJAMENA
(Fort Lamy)

1985
NIGER
CHAD
NIGERIA
CAMEROON
Chari
NDJAMENA
(Fort Lamy)

2000
NIGER
CHAD
NIGERIA
CAMEROON
Chari
NDJAMENA
(Fort Lamy)

Rainfall variation in the Sahel 1922–1994

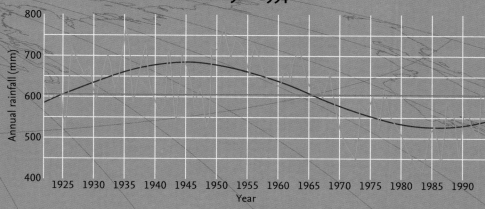

Annual rainfall (mm) — Year

Annual total rainfall
Long-term average
72-year cyclical pattern

Water use in the Yellow River basin 2000 and 2010 (projected) (x10^8 m^3)

Region	Year	Industrial	Urban residential	Rural residential	Agricultural	Region total 2000	Region total 2010
Upper reaches	2000	23.2	4.5	4.9	187.9	220.5	
	2010	36.2	6.3	5.9	190		238.4
Middle reaches	2000	77.5	9.4	11.5	117.6	216	
	2010	111.3	14.5	14	127.2		267
Lower reaches	2000	27.4	4.6	2.8	167.1	201.9	
	2010	41	6.5	3.4	167.1		218
Category total	2000	128.1	18.5	19.2	472.6	**638.4**	
Category total	2010	188.5	27.3	23.3	484.3		**723.4**

No-flow days of Yellow River

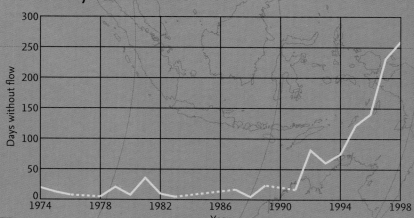

Days without flow — Year

A similar fate is befalling Lake Chad in the Sahel region of Africa. This lake is now down to 5 per cent of its 1960s size, thanks to a combination of river diversions and perennial drought. In Mexico, the country's largest body of water, Lake Chapala, has been in long-term decline since the 1970s, and lost 226 sq km (87 sq miles) in the last five years alone. In China's Yangtze river valley, more than 800 lakes have disappeared entirely in the last half century. All told, more than half the world's five million lakes are endangered.

As well as providing fisheries and fresh water, lakes also provide flood protection by storing excess water and releasing it slowly into rivers. They are also vital parts of the ecosystem, supporting wetland and migrating birds, diverse fish species and other aquatic life. A disappearing lake will often also mean a disappearing animal or plant species, contributing to a global reduction in biodiversity.

Lake Chad, Africa 1972

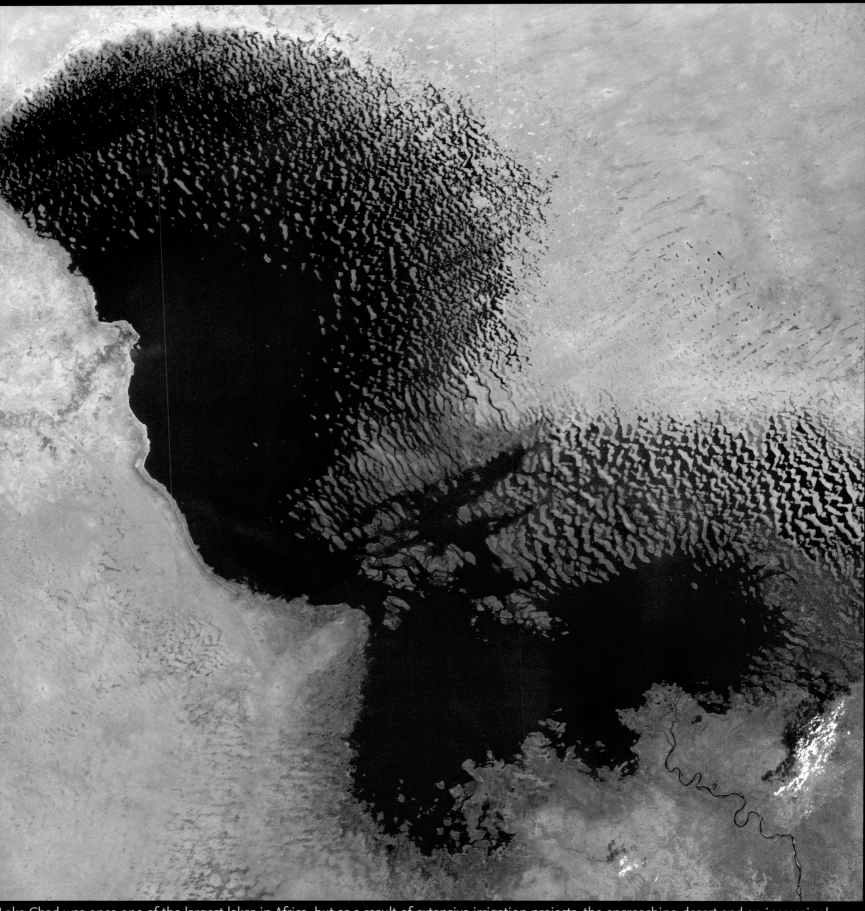

Lake Chad was once one of the largest lakes in Africa, but as a result of extensive irrigation projects, the encroaching desert and an increasingly dry climate, it is now a twentieth of its former size. As the lake floor is flat and shallow, the water level fluctuates seasonally with the rainfall. The fifteen years from 1972 to 1987 saw the most dramatic change in the lake, as illustrated in these two satellite images.

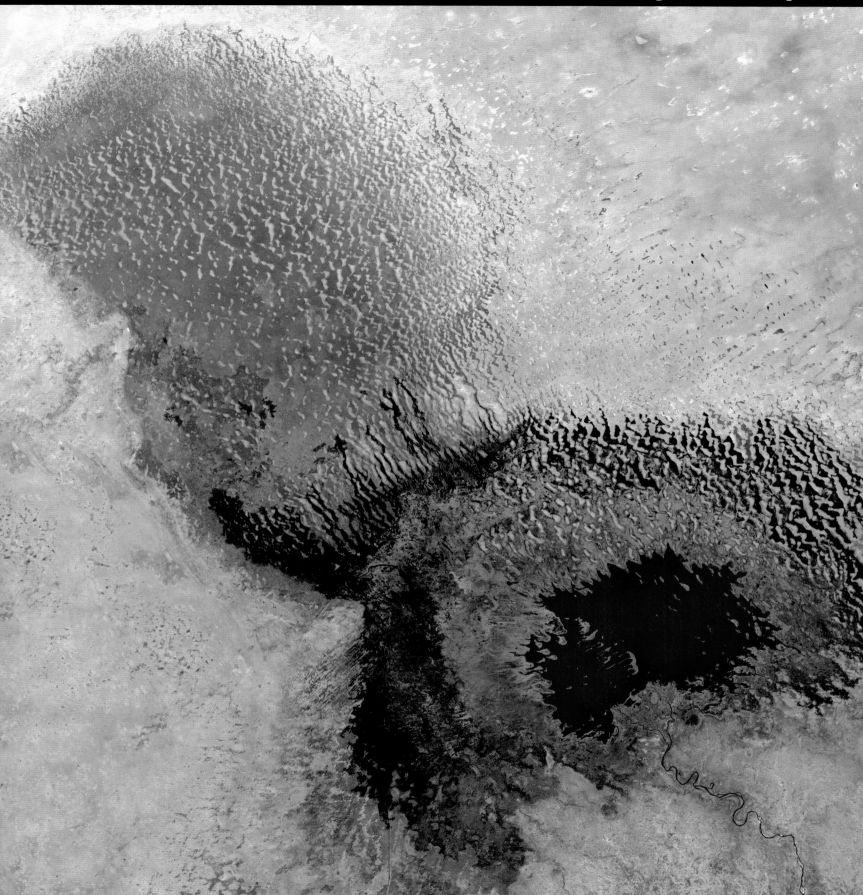

This dramatic change was the result of an increase in water being diverted for irrigation. Now, with a drying climate, the desert is taking over, as is shown by the ripples of wind-formed sand dunes where the northern half of the lake used to be.

A dying lake, Aral Sea, Central Asia

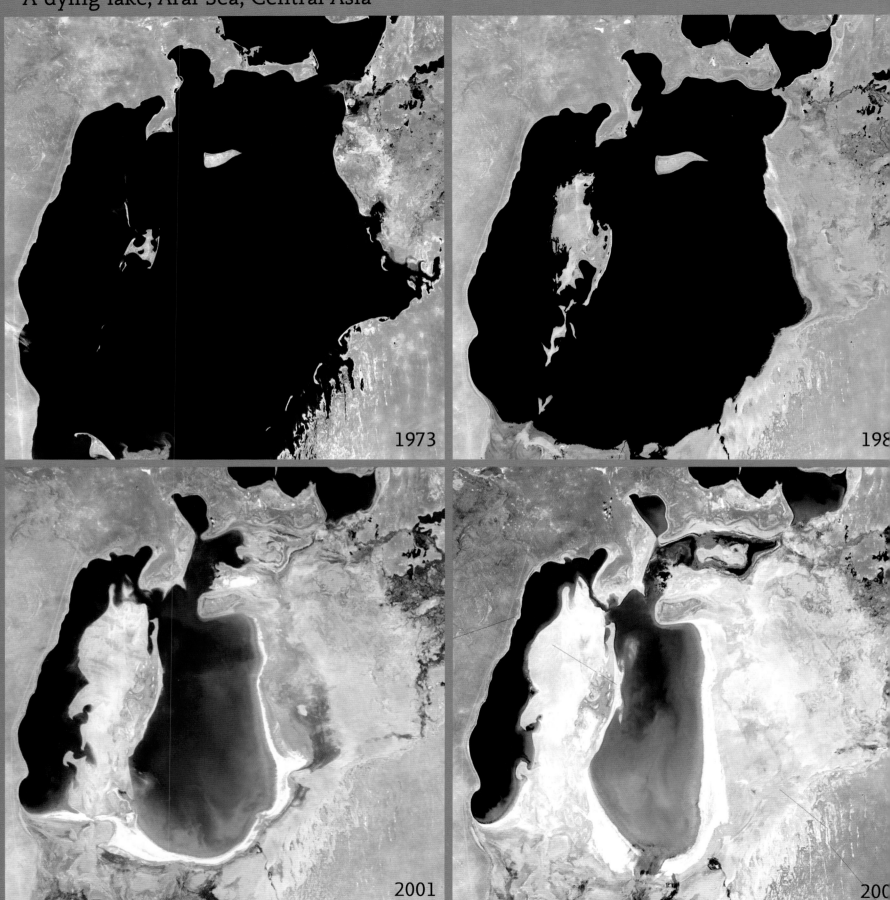

1973

198

2001

200

The Aral Sea was once the world's fourth-largest lake. Today, due to climate change and the diversion of water from its feeder rivers for irrigation, it is much smaller. Steps are being taken to preserve the northern part by constructing a dam, but the southern part has been abandoned to its fate.

The local fishing industry on the Aral Sea has been devastated by the lake's shrinkage and the local population has developed health problems due to the exposure of chemicals on the dry sea bed. Abandoned ships litter the former lake bed and as it dries out vast salt plains are forming and dust storms are becoming more frequent.

Udaipur, known as the City of Lakes, has many Rajput palaces. The most famous of these is the Lake Palace, built entirely of marble on a small island in Pichola Lake. Operating as a luxury hotel, until recently it was accessible only by boat.

As a result of severe drought in Rajasthan and seepage of water, the lake recently dried out. This photograph shows the hotel high and dry and having lost some of its romantic appeal. It is now accessible by road.

Colorado River, USA 2002 and 2003

These two images of the Colorado River as it enters Lake Powell illustrate the effect of low rainfall in the western USA. The Lake Powell reservoir level had fallen by 13 m (43 feet) in the eighteen months between photos. By 2005 the lake was at its lowest ever level.

Drought is not restricted to countries with a hot, dry climate. In 2003, and again in 2004, dry and hot summers resulted in the water level of many rivers in western Europe falling to very low levels. This hindered navigation on major rivers such as the Elbe and Rhine in Germany.

Water's

Power

Coast and flood

Changing coastlines – *the action of the sea and rivers in the erosion of coastal features and the deposition of sediment along the coast*

Rivers in flood – *water inundating normally dry land as river levels rise after heavy rainfall or as a result of melting snow*

The formation of Dungeness, Kent, UK

Romney Marsh

New Romney

Rother

Walland Marsh

Lydd

Rye

Dungeness

Erosion

Longshore drift

4

3

2 Suggested former shorelines

1

Land enclosed during Roman times

Enclosed 1100–1300s

Enclosed 1400–1600s

Ancient cliffs

The Earth's coastlines may look permanent to the casual observer, but in reality they are always changing. Land is constantly destroyed by the sea as waves and currents erode beaches and cliffs, but new land is created as sediments are re-deposited by ocean currents elsewhere. Dry land is also formed when rivers create deltas by washing silt out to sea, and – over geological timescales – by tectonic uplift. Volcanoes can create new land very rapidly when they erupt into the sea: lava flows from the eruption of Montagu in the South Sandwich islands, near Antarctica, created 20 hectares (50 acres) of new British Overseas Territory at the end of 2005.

Holderness coastal erosion, UK

Direction of movement of sediment

Roman coastline
Projected coastline 2100
'lost' villages

Holderness coastline at risk – one of the fastest eroding coastlines in Europe

Coastal distance lost in 2000 years	400m (437yds)
Average rate of erosion	2m (6.5ft) per year
Largest loss recorded	6m (19.7m) in 2 days (Barmston, October 1967)
Material lost in last 100 years	76 450 000 cubic m (99 992 819 cubic yards)
Deposition at Spurn Head	3% of material eroded
Other deposition areas	Deep water offshore, in Humber Estuary, North Lincolnshire coast
Other losses	Over 30 village sites

Changing coastlines

Coastal erosion rates

Rock type	Erosion rate	
	mm per year	inches per year
Granite	0.1	0.004
Limestone	0.1–1.0	0.004–0.04
Shale	1	0.04
Chalk	1.0–100	0.04–3.94
Sandstone	1.0–100	0.04–3.94
Glacial till	100–1000	3.94–39.37

Moving islands – eastern USA

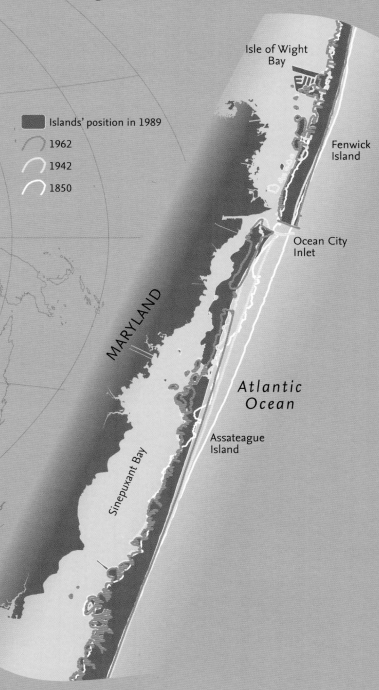

Islands' position in 1989

1962

1942

1850

Different coastlines erode at different rates according to their constituent materials. Tough rocky cliffs may be relatively impervious to the daily scouring of waves and currents, but mud and clay coastlines erode relatively quickly. The eastern coast of England has been retreating for millennia: the Suffolk town of Dunwich became a 'rotten borough' by the eighteenth century, electing two MPs to Parliament in London even after almost the entire town had been washed into the sea.

Climate change has a direct impact on the coast by driving up global sea levels. Over 70 per cent of the world's sandy shorelines are currently retreating due to sea level rise. Global warming also affects coastal erosion by giving the waves more power as storms get stronger. This is a particular concern for hurricane-prone regions of the world such as the Caribbean and the east coast of the USA.

In the Arctic, coastal erosion is a relatively new problem, because shorelines which used to be in the grip of permafrost and sea ice have now begun to melt and collapse. The Alaskan village of Shishmaref is planning to move off its rapidly-eroding barrier island because melting sea ice has allowed waves to undercut its sandy cliffs (see pages 162–163).

The 1950s wooden sea defences at Happisburgh are failing and large chunks of cliff regularly fall into the sea. This part of the Norfolk coastline is soft and vulnerable to coastal storms. As seen in this pair of photographs, several buildings have vanished over a period of four years with homes, businesses, roads and the lifeboat station all at risk.

The west coast of the USA is vulnerable to storm damage. These photographs were taken as part of a project to record coastal erosion resulting from severe storms generated as part of the Pacific Ocean warming effect of El Niño. In just one winter a significant amount of

The Twelve Apostles are famous Australian coastal landmarks, which were created by the sea gradually eroding soft limestone cliffs. Cliffs are eroded into headlands, in which caves then form. When caves break through a headland a sea arch is formed, then when an arch collapses a

The Twelve Apostles are now only eight in number. These two images were taken less than one minute apart on a Sunday morning, when one of the 50 m (164 ft) high stacks gave way to the power of the sea and collapsed into a pile of rubble, much to the shock of onlookers.

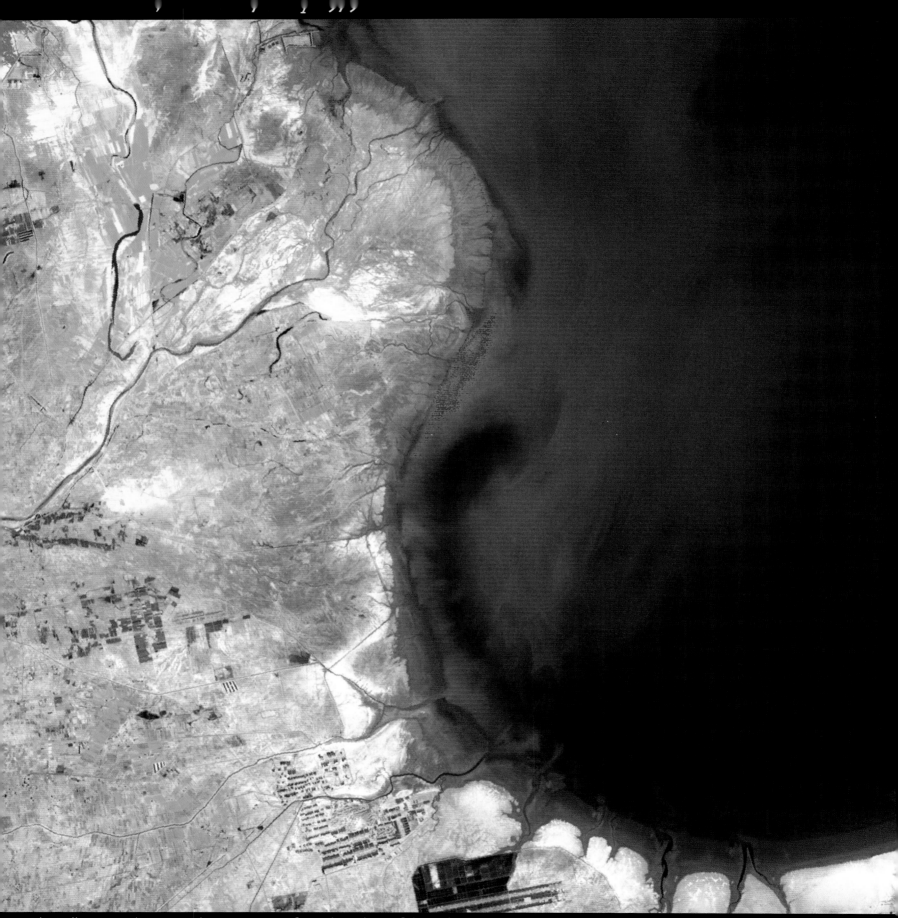

The Yellow River (Huang He) gets its name from the colour of the sediment it carries. This is mainly mica, quartz and feldspar and as the river travels through north central China it crosses an easily eroded loess plateau. Loess is called huang tu, or 'yellow earth' in Chinese, and is

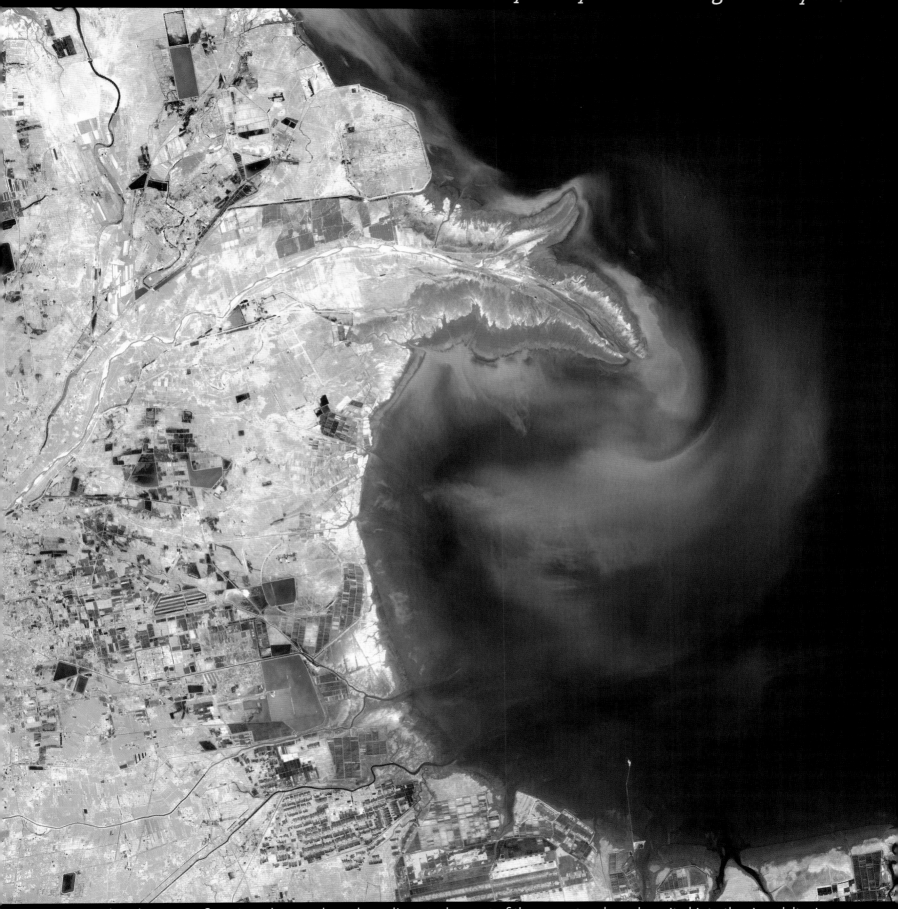

Once the river reaches the coast it flows into the sea where the sediments drop out of the current and are deposited into the river delta. In

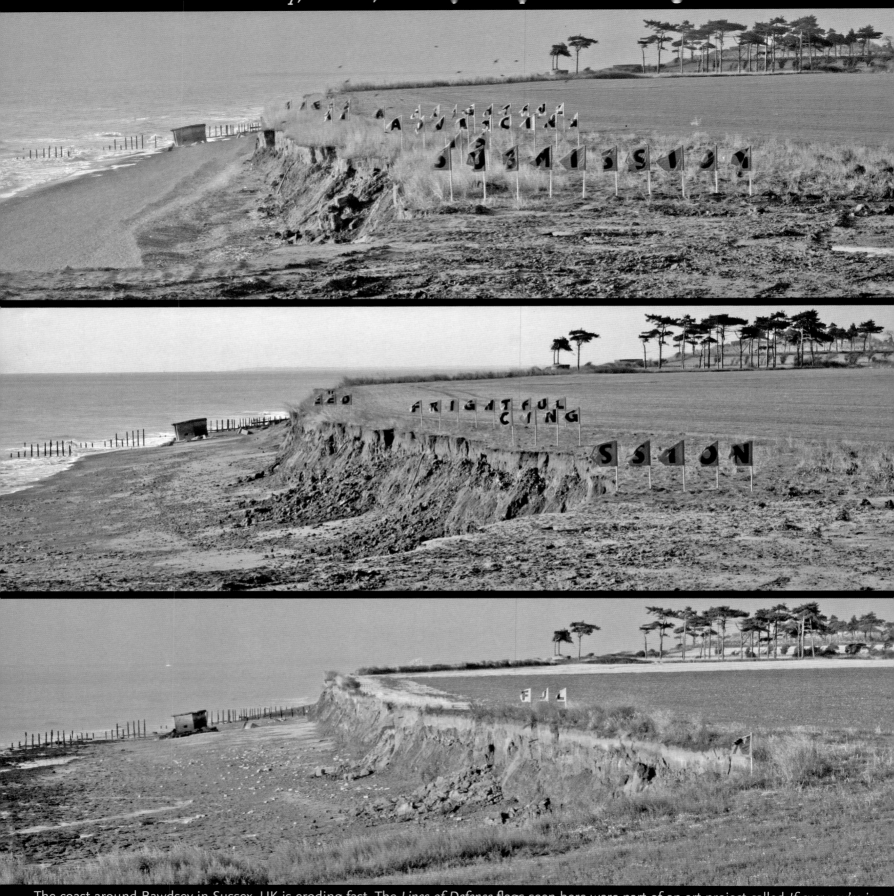

The coast around Bawdsey in Sussex, UK is eroding fast. The *Lines of Defence* flags seen here were part of an art project called *If ever you're in the area*. The complete set of project images beautifully illustrate the problem as they cover a full year. These three images illustrate the erosion from January to August 2005, by which time approximately 14 m (46 feet) of coastline had been lost.

This 7 km (4 miles) of raised shingle bar is unusual in the UK, where many coastal features of this type have been built on or removed. Culbin Bar is a virtually untouched remnant of a much larger area of sand and shingle coastline. It seems to have started out as a spit growing from the eastern shore of the river Findhorn. The bar changes shape due to erosion at its eastern end and active deposition at its

Cape Hatteras is part of a barrier island system on the coast of North Carolina. As a result of rising sea levels, strong storms and warming of

By 2004 we can see how much the coastline of this important tourist area has moved in just five years, leaving this and many other houses nearby close to destruction. Insurance money is often used for building large beach houses which are rented out to visitors, rather than for replacing small motels and beach cottages which were more usual in the past.

River floods can be hugely destructive to life and property, but they play an essential role too. In Bangladesh, images of families forced from their homes by flooding are familiar, but the rising waters are a vital part of an annual cycle – bringing fertility to the fields and replenishing groundwater supplies. Until the building of the Aswan Dam in Egypt, yearly floods brought nutrients to the agricultural land along the river bank. Since the dam began operating in 1964, farmers have had to resort to artificial fertilizers, and parts of the Nile Delta have begun to sink because of the reduction in sediment flowing downstream.

Human interference with rivers makes it very difficult to give an accurate picture of how floods are changing in frequency and magnitude. Recent decades have seen big increases in flood damage, but much of this could be the result of bigger cities and more built-up areas expanding into river flood plains. The construction of levees can help reduce the damage, but if these dykes are breached they can actually work in reverse, preventing water flowing back into the river as it subsides, and making the flooding worse.

Rivers in flood

Total annual precipitation

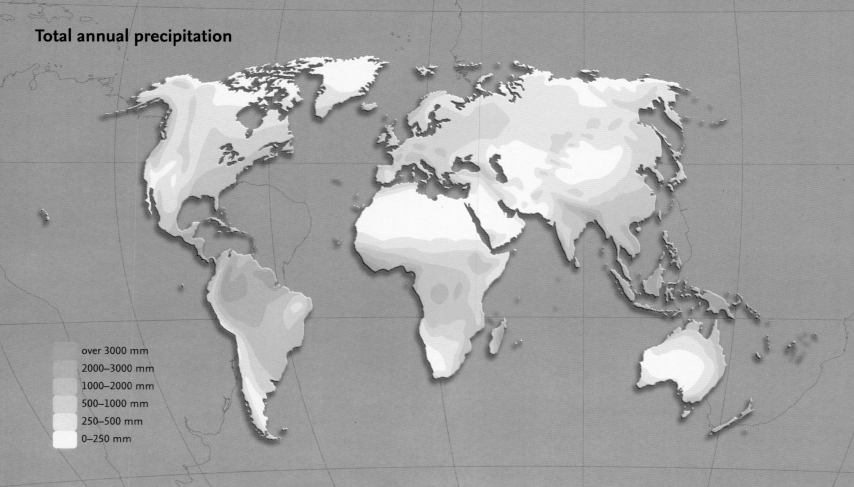

over 3000 mm
2000–3000 mm
1000–2000 mm
500–1000 mm
250–500 mm
0–250 mm

Flooding has a clear relationship with precipitation intensity. Heavier rain and snowfall are undoubtedly linked to a changing climate because a warmer atmosphere can hold more water vapour. Global warming is already making rainfall heavier across the world, from the United States to Japan. In the UK the last thirty years have seen a doubling of rainfall events in the heaviest category, contributing to the devastating floods in October/November 2000. This trend is predicted to continue in the future, with more intense summer monsoons bringing worse flooding to Bangladesh, India and southern China, and heavier winter rainfall affecting higher latitude areas in Europe and North America.

Significant precipitation events 2004

1. Severe flooding in April in the Escondido river area Mexico/USA.
2. Southern USA wetter than average.
3. Eastern USA wetter than average.
4. Severe cold and snow during June and July in Peru.
5. During a wet December to February, Brazil's northeastern states suffer severe January flooding.
6. Severe cold and snow during June and July in Chile and Argentina.
7. Russia and Belarus affected by flooding from March until May with over 1000 people displaced.
8. Widespread winter storms in February over Syria, Greece and Turkey with around 600mm of snow in Jordan.
9. Severe floods in April in Angola and they continue into May in Zambia.
10. Monsoon-related floods from June to October in India and Bangladesh, the worst floods for 15 years with many millions displaced.
11. September floods in China kill 196 people.
12. Japan was wetter than average all year, partly due to tropical storms making landfall.
13. Northern Australia much wetter than the average rainy season.
14. Heavy rain and damaging floods in New Zealand during February.

Deadliest floods since 1900

Date	Country	Location	Estimated death toll
1911	China	Yangtze	100 000
1931	China	Yangtze	3 700 000
1933	China	Henan, Hebei, Shandong, Jiangsu	18 000
1935	China	East	142 000
1939	China	Henan	500 000
1949	China	Northeast	57 000
1949	Guatemala	East	40 000
1954	China	Yangtze	30 000
1959	China	North	2 000 000
1960	Bangladesh	Central	10 000
1974	Bangladesh	Dhaka	28 700
1999	Venezuela	North	30 000

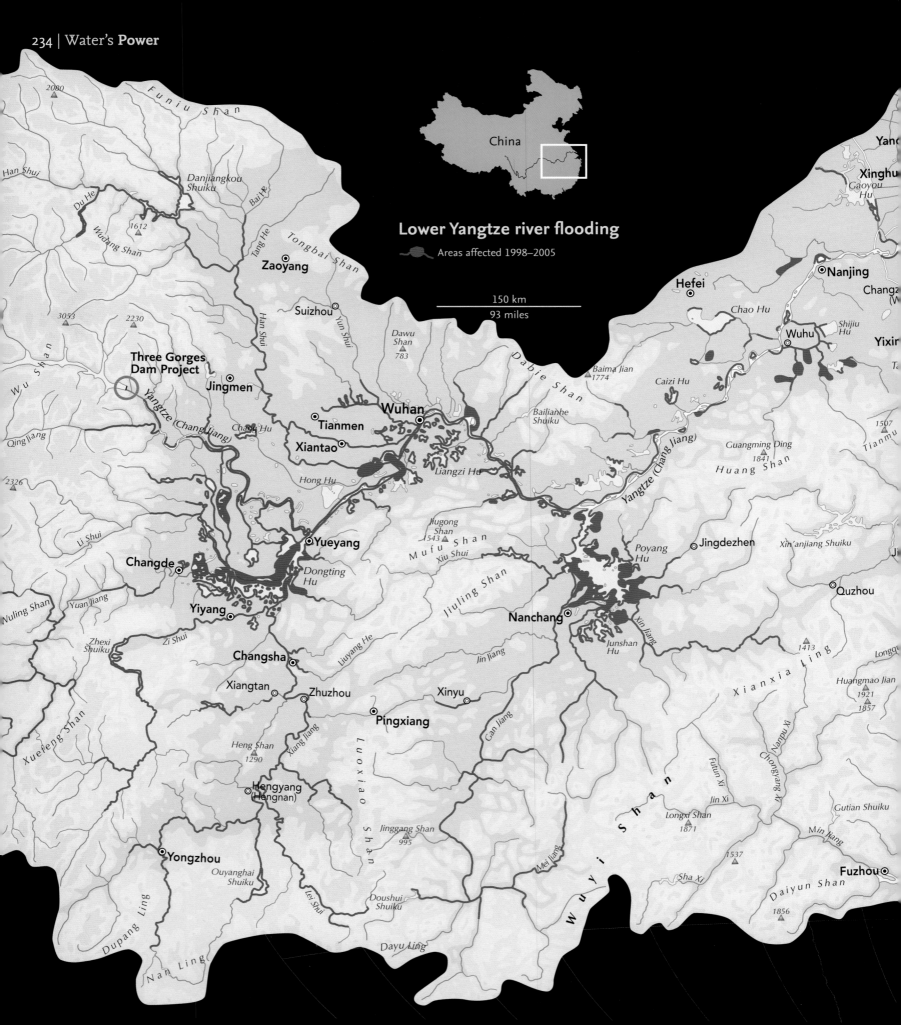

China

Lower Yangtze river flooding

Areas affected 1998–2005

150 km
93 miles

Three Gorges
Dam Project

The Yangtze basin is home to 400 million people, making it one of the most densely populated river basins in the world. It is also one of the most flood-prone areas, and hundreds of thousands of people died in floods there earlier in the twentieth-century. In the top-ten list of the world's deadliest floods since 1900, seven have been in China. Rainfall in the country is highly variable, with the northern half frequently subjected to drought, while the southern half receives summer deluges. Flooding tends to be worse in El Niño years, when changing ocean currents in the tropical Pacific spread weather chaos around the globe.

China's relationship to world flood statistics

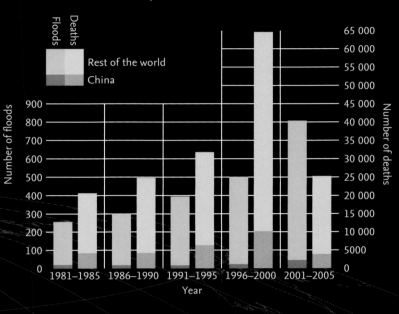

The Chinese have built an elaborate system of dykes, dams and flood gates in order to try and tame the Yangtze, and to generate electricity. The Three Gorges Dam – the biggest hydroelectric dam in the world, with a reservoir capacity of 22 billion cubic metres – is intended to reduce the frequency of downstream flooding by ten times as well as to generate a tenth of the entire country's electricity. Overall, the Yangtze is much more of a boon than a threat: it is one of China's busiest waterways for 1000 km (621 miles) along its length, and its waters irrigate rice fields in Jiangsu province helping to keep the country self-sufficient in food.

But severe floods in July 1998 illustrated the threat the Yangtze can still pose. The floods were caused primarily by exceptionally heavy rainfall, but were exacerbated by deforestation and over-development. Forests help to mitigate floods by storing water and releasing it slowly, whereas denuded slopes can see catastrophic flash floods. 85 per cent of the Yangtze basin's original forest cover has now been cut down. However, the floods were in some senses a success story for the government: even with 250 million people affected, the death toll was kept below 5000.

Rainfall in the Yangtze basin
Rainfall in 1998 and 2002 compared to normal

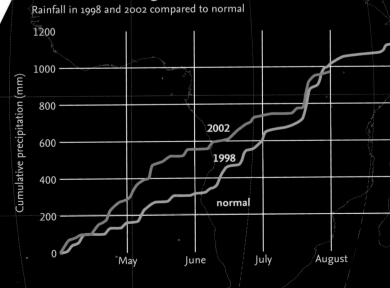

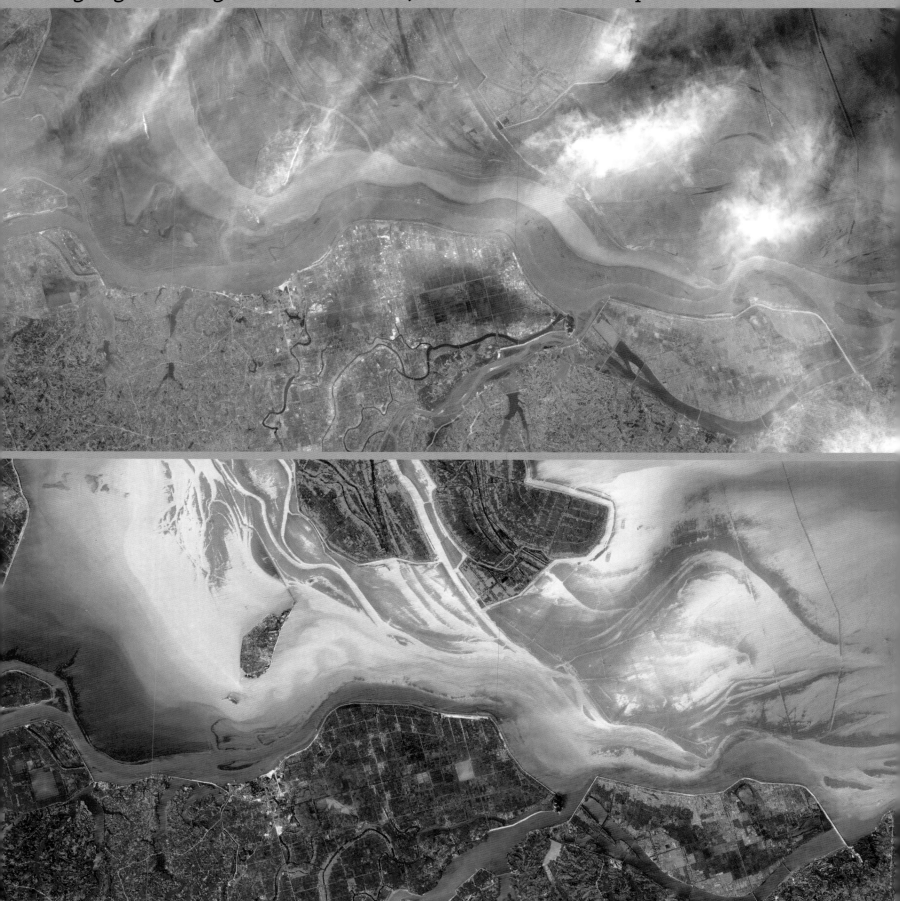

Dongting Hu, Yangtze River, China 19 March 2002 and 2 September 2002

Dongting Hu is a lake on the Yangtze river. (See page 234). It is the second largest freshwater lake in China, but is only approximately 45 per cent of its size of 150 years ago when it covered 6200 sq km (2394 sq miles). Land reclamation has reduced its size, but this activity was stopped in 1981. However, the high sediment load in the Yangtze continues to be a huge problem, silting up the lake.

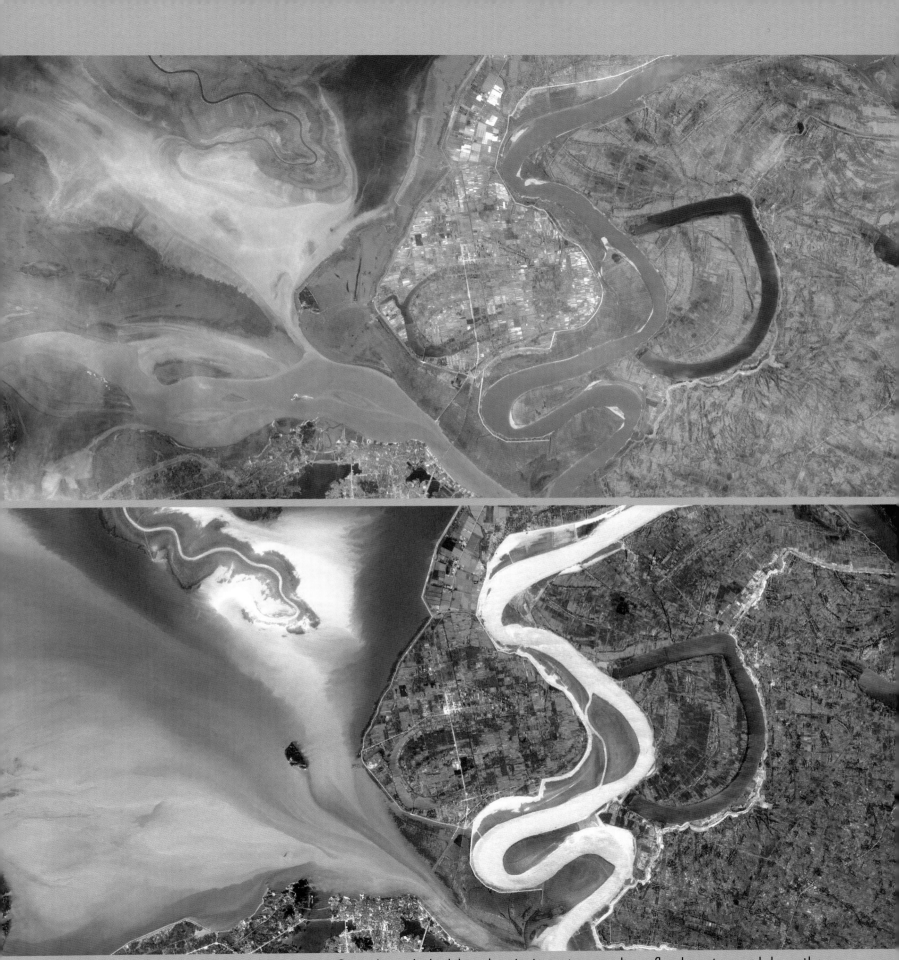

As approximately 40 per cent of the Yangtze river water flows through the lake, when in August 2002 a large flood crest surged down the river inundating the lake area, as seen in the lower image it was potentially catastrophic for the local people. Fortunately, the embankments made by the locals held up, but in the past flood waters have inundated the land around the lake.

Betsiboka River, Madagascar 4 September 2003 and 25 March 2004

Catastrophic erosion in northwestern Madagascar has resulted from the removal of native forest for timber. The top image shows normal river levels but below that the widespread flooding and massive red sediment plume as a result of tropical cyclone Gafilo, which hit northern Madagascar on 7 and 8 March 2004 can be seen. Not only is the soil upstream eroding but the sediment is silting up the estuary causing further problems.

Flooding in Western Australia 20 February 2006 and 2 March 2006

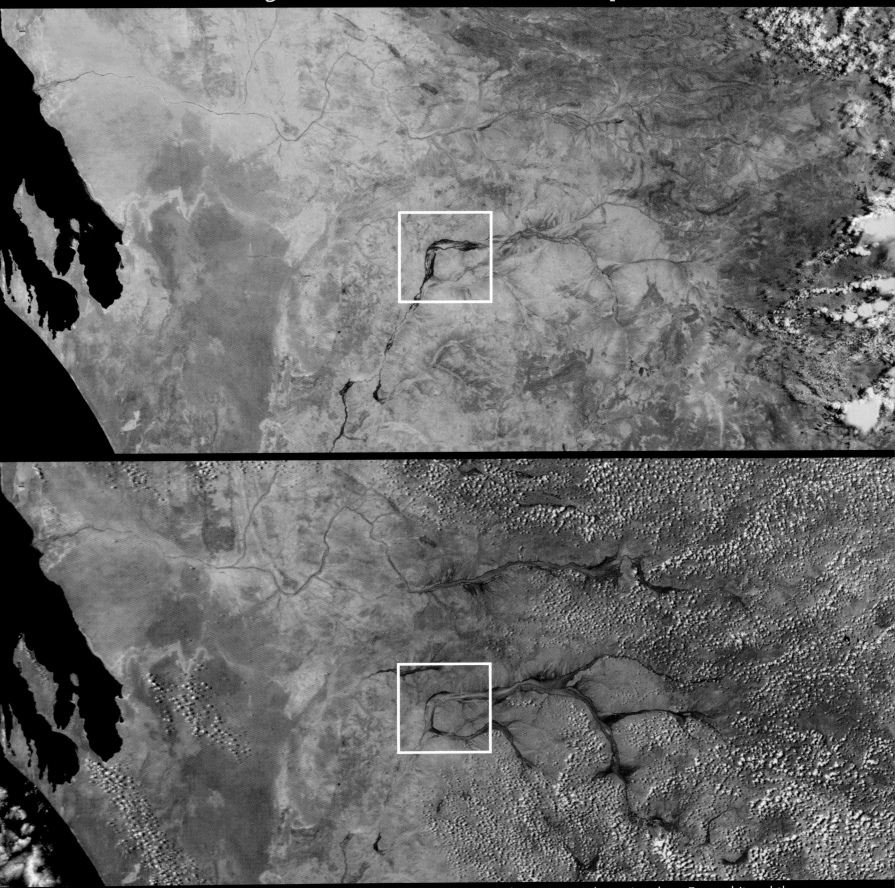

A typical late summer season view of Western Australia can be seen in the top image, but on 1 March 2006 cyclone Emma hit and the ground was deluged with heavy rain. This triggered widespread flooding in the Murchison and Gascoyne river basins as seen in the bottom image. Vegetation has flourished in this wet period, making the flooded rivers stand out clearly.

Until relatively recently, whenever the River Severn burst its banks and flooded, Bewdley suffered badly, as can be seen in the top image. However, flood defences have now been installed. During a flood a defence wall is now raised, keeping the buildings behind safe.

Kloster Weltenburg, Bavaria, Germany 31 July 1999 and 25 August 2005

Kloster Weltenburg, a normally serene Benedictine monastery on the banks of the Danube, dates back to AD 600. There has been a brewery on the site since 1050 and it is a popular tourist destination. In August 2005 the river rose by 7 m (24 feet), flooding the monastery buildings, but not the brewery.

In the area around St Louis, the Mississippi, Missouri and Illinois rivers all meet. The land cover along the river banks has been changed from natural vegetation to agricultural land or is built up. This means that the wetlands which can absorb large amounts of water and release it slowly over time are missing, replaced by levees, canals and dams.

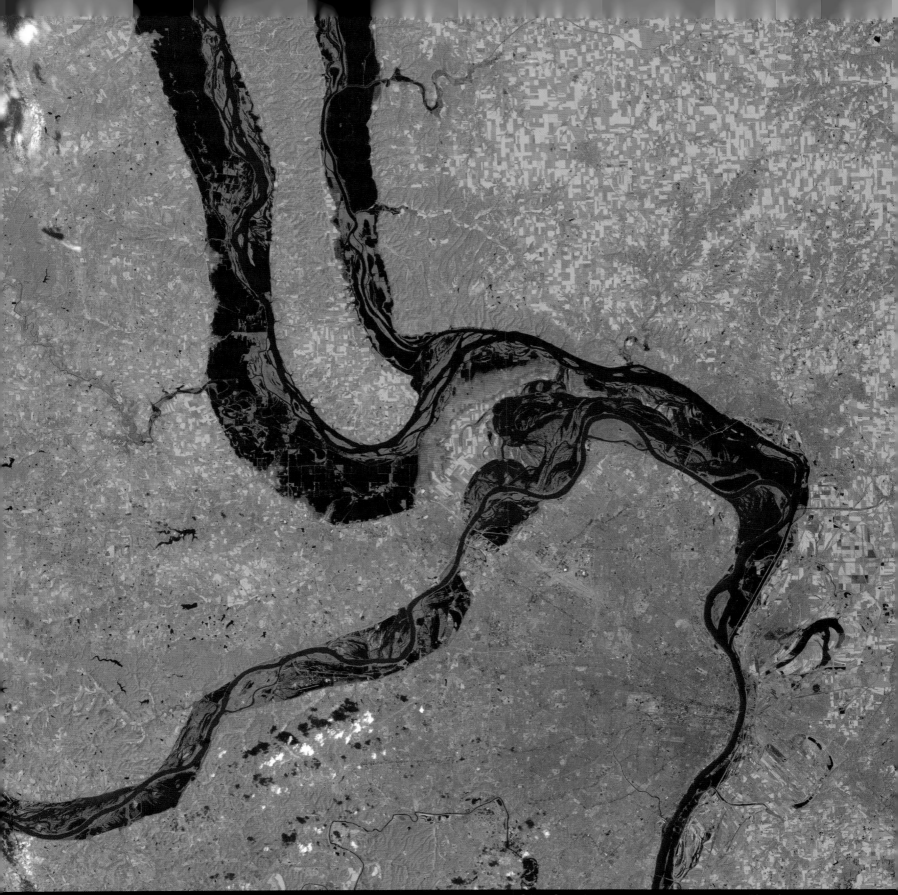

In early 1993 the upper Mississippi drainage basin received up to twice the average rainfall, often during very intense storms. Floods overwhelmed the water control structures in the basin leading to the biggest flood ever recorded in the area. In St Louis the Mississippi stayed above flood stage from 1 April until 30 September. In this August image the pink areas show exposed soil as the water

Future

Views

Changing world

A view into the future
Elizabeth Kolbert

A few years ago, in an essay in the journal *Nature*, the Nobel Prize-winning Dutch chemist Paul Crutzen coined a term. No longer, he wrote, should we think of ourselves as living in the Holocene, as the ten thousand years since the last glaciation is known. Instead, an epoch unlike any of those which preceded it had begun. This new age was defined by one creature – man – who had become so dominant that he was capable of altering the planet on a geological scale. Crutzen dubbed this age the Anthropocene.

Crutzen's was not the first such neologism. Already in the 1870s, the Italian geologist Antonio Stoppani had argued that human influence was ushering in a new age, which he called the 'anthropozoic era'. A few decades later, the Russian geochemist Vladimir Ivanovich Vernadsky proposed that the Earth was entering a new stage – the 'noosphere' – dominated by human thought. But while these earlier terms had had a positive slant – 'I look forward with great optimism… We live in a transition to the noosphere', Vernadsky wrote – the connotations of the Anthropocene were distinctly cautionary. Humans had become the driving agents on the planet, yet, according to Crutzen, they didn't seem to have given much thought to where they were heading.

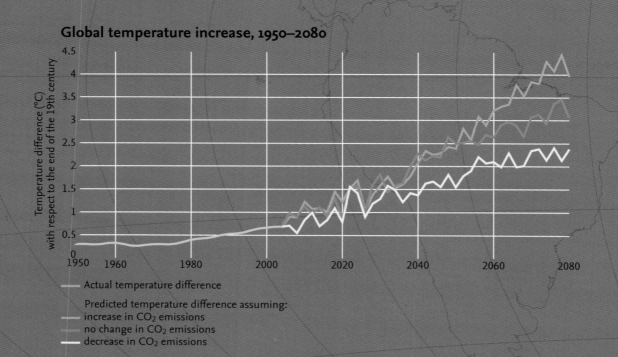

Global temperature increase, 1950–2080

Temperature difference (°C) with respect to the end of the 19th century

—— Actual temperature difference

Predicted temperature difference assuming:
—— increase in CO_2 emissions
—— no change in CO_2 emissions
—— decrease in CO_2 emissions

Thanks to some of the same technologies which have allowed humans to dominate the planet, we now know a great deal about the future of the Anthropocene. For a start, we know that the planet will grow warmer. The climate operates on a tremendous time-lag, a result of what scientists call 'thermal inertia'. The greenhouse gases which we have already added to the atmosphere have, in effect, turned up the Earth's thermostat, and because these gases are so persistent, there is nothing much we can do at this point to lower it again. Even if greenhouse gases could somehow magically be held constant at today's levels – and this would require some pretty heavy-duty magic – global temperatures would continue to climb at least through the middle of this century.

We also know that as a result of this warming, sea levels will rise. Water expands as it heats up. In a small body of water, the effect is small; in a big body, it is commensurately larger. Most of the sea level rise which has been predicted for the next hundred years is purely a function of thermal expansion. The latest data suggest that, over the last few years, both Greenland and Antarctica have been losing ice at an alarming rate. If these losses continue, then sea level rise will be significantly more dramatic than previously predicted.

 In addition to warming and expanding, the oceans will become more acidic. The seas absorb much of the carbon dioxide which we emit from our cars and power plants, and this dissolved carbon dioxide forms an acid which alters the chemical properties of the water. As carbon dioxide levels in the air continue to increase, the acidity of the seas will increase. Organisms that form shells of calcium carbonate, which dissolves in acid, much like chalk in a bowlful of vinegar, are apt to be particularly hard hit by the change. The Royal Society has concluded that the acidification of the oceans which has already occurred owing to human emissions 'will take tens of thousands of years' to reverse.

Higher temperatures, rising seas, acidifying oceans – these changes can be predicted with something very close to certainty. What is less certain is how large, which is to say how catastrophic, they will be. In part, the magnitude of the changes will depend on natural feedbacks which, at this point, are not completely understood. For instance, warm water takes up less carbon dioxide than cold water, so as the seas heat up absorption of the gas by the oceans is likely to decline, meaning that more will remain in the atmosphere, contributing to further warming, and so on.

In part, of course, the magnitude of the changes depends on us. Humans aren't the first species to alter the planet on a geological scale; that distinction probably belongs to early bacteria, which, some two billion years ago, invented photosynthesis. But we are the first species to be in a position to understand what we are doing.

The future of the world
Bjørn Lomborg

When looking at the dramatic pictures in this book, the future seems daunting. But we have to remember that humanity has always lived at the mercy of the elements – endangered by diseases, predators and the fury of natural forces. If anything, our history shows that over the past centuries we have managed to gain more control and security. Humanity's lot has improved dramatically – both in the developed, where it is rather obvious, but also in the developing world, where life expectancy has more than doubled over the past hundred years; malnutrition has dropped from 50 per cent in 1950 to 17 per cent today; poverty has fallen from 50 per cent to 25 per cent; access to clean drinking water has gone up from 30 per cent in 1970 to 80 per cent today; and illiteracy has dropped from 80 per cent to 20 per cent.

In the rich world, the environmental situation has also improved. In the United States, the most important environmental indicator, particulate air pollution, has been cut by more than half since 1955, rivers and coastal waters have dramatically improved, and forests are increasing in size. And these trends are generally shared by all developed countries. Why? Because we are now rich enough to care for the environment. In much of the developing world, environmental conditions are getting worse. But these countries are only acting as we once did. They care first about feeding their kids before cleaning up the air. Affluence will make the environment a higher priority. In some of today's richer developing countries, such as Mexico and Chile, air pollution is already beginning to decline.

Although we constantly fear that the future will overwhelm us, this often comes from faulty analysis. We believe that we will be flooded in our own trash. Take the USA, where ever more people throw out ever more trash. Over the next one hundred years, all this trash will not run out of control. If it was all to be placed in one location, and kept to a height of 30.5 m (100 feet), it would only take up a square measuring 29 km (18 miles) on each side. Really this is just a logistical problem.

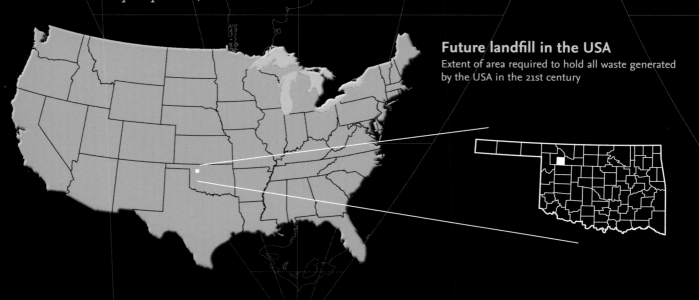

Future landfill in the USA
Extent of area required to hold all waste generated by the USA in the 21st century

We worry about the seemingly ever-increasing number of natural catastrophes. Yet this is mainly a consequence of media news coverage. We see many more disasters, but the number is roughly constant and we manage to deal much better with them. Globally, the death rate from catastrophes has dropped about 50-fold over the past century, mainly due to better technology and wealth. In the 1970s, Bangladesh routinely saw hundreds of thousands die from flooding; today warning systems and shelters have dramatically reduced the death toll. We also worry that global warming will increase flooding and hurricane damage. And yes, sea levels will increase 0.3–0.5 m (1–1.6 feet) over the coming century. But we must remember that during the last century sea levels rose 10–25 cm (4–10 inches), and nobody noticed. We adapted. The most pessimistic United Nations predictions expect the average Bangladeshi in 2100 to be as rich as we are today. Bangladesh will then be able to deal with rising tides while protecting its population even better.

The jury is still out on whether global warming causes increasing hurricane intensity. But we worry about the wrong issue. Hurricane damage is increasing, but predominantly due to more people with more goods settling in ever more perilous locations. Even if global warming is increasing the strength of hurricanes, it will only cause 2–5 per cent of the future damage. Should we not primarily be concerned about the other 95–98 per cent? Captivating and scary pictures are good at focusing our attention, but we have

to remember that they don't necessarily show us the most important issues or the best places to intervene. Silent and dispersed problems are often much more important. Poverty leaves you in danger from disease and catastrophe, and makes it hard to care for the environment.

How do we make a better world? This question was answered by the Copenhagen Consensus project. Here, eight of the world's top economists (including four Nobel Laureates) established a global priority list based on elaborate assessments by thirty specialist economists. At the top of the list they put preventing HIV/AIDS, malnutrition, and malaria, along with ending agricultural subsidies. These are the areas in which we can do the most good per dollar for the world. Kyoto ended up at the bottom of their list, because it would cost a great deal and do little good. Why did thousands die in Haiti during the recent hurricanes and not in Florida? Because Haitians are poor. They cannot take preventive measures. Breaking the circle of poverty, by addressing the most pressing issues of disease, hunger and trade opportunities, will not only do obvious good but will also make people less vulnerable to the effects of climate change.

It is likely that the future of the world will be better – simply because our technology and innovation keep making us richer and better able to deal with our threats. The trick is to worry about the right things first.

Looking to the future
Mark Lynas

For the first time ever in human history, we have the capacity today to accurately predict the future. In the past, soothsayers consulted oracles, made sacrifices or read tea leaves. Today, scientists construct complex computer models which are driven by hundreds of equations representing the real laws of physics in action: how the atmosphere circulates, how water evaporates and precipitates, how solar radiation is reflected or retained by different surfaces and so on. When run on huge supercomputers the size of several tennis courts, these virtual models of the planet can give us significant insights into how changes in climate are likely to unfold as greenhouse gases accumulate in the atmosphere.

The story they tell is a sobering one. Scientists predict that by 2100 global temperatures will have risen by between 1.4 and 5.8 degrees Celsius. The lower end of this scale would cause significant disruption for ecosystems and human society. Mountain glaciers will continue to melt, tropical coral reefs could disappear almost completely, while stronger storms and droughts will place increasing pressure on the world's economy. Sea levels will rise by 0.5 m (1.6 feet) or so, displacing people from low-lying islands and deltaic coasts. The upper end of this scale, however, promises destruction on a scale never experienced before by human civilization, and some commentators have suggested that society itself could collapse. A geological perspective helps put the magnitude of these changes into context: a rise in planetary temperature of five degrees would make it hotter than for over fifty million years, well before humans began to evolve.

Bangladesh at risk

Land flooded if sea level were to rise by 7 m (23 feet)

Computer models give us some very specific ideas of what lies in store at different degrees of temperature change. Two degrees is the threshold above which the Greenland ice sheet will melt irreversibly, according to several studies. This would eventually raise global sea levels by 7 m (23 feet), flooding coastal cities such as London, New York and Bangkok, as well as half of Bangladesh. If western Antarctica too begins to melt rapidly, the world's coastlines will begin to look very different. Major uncertainties remain about how quickly this might happen. Most glaciologists think that Greenland would take several centuries to melt completely, although NASA climatologist James Hansen has suggested that the process might happen much more quickly.

According to a model produced by the Hadley Centre, part of the UK's Meteorological Office, at an increase of three degrees the Amazon rainforest ecosystem would begin to collapse. Fires would sweep through the river basin as rainfall totals plummeted, converting what is now lush forest into savanna or even desert. Given that the Amazon is home to over half the world's species, the ecosystem's collapse would be a disaster for biodiversity, contributing to a planetary mass extinction. A separate modelling study by ecologists has suggested that a million of the world's species could be doomed to extinction before 2050 by this amount of warming.

There are lessons from the past, too. Geological studies of the mass extinction which took place at the end of the Permian period, 251 million years ago, have suggested that a sudden bout of global warming — driven, perhaps, by volcanic eruptions and a sudden burst of methane from the oceans — of around six degrees wiped out up to 95 percent of species which were alive at the time. Since this is the top range of scientific projections for 2100, this outcome is unlikely, but extremely worrying nonetheless.

Ultimately, of course, the future has yet to be written, and the most pessimistic computer model predictions need never become reality if human society decides to reduce the emissions of greenhouse gases to sustainable levels. This is the biggest uncertainty of all, because it depends on decisions which have yet to be made by billions of people, decisions which are mundane and everyday (whether to drive or cycle, for example) but which, when taken together, will ultimately determine the fate of our fragile planet over the century ahead.

Water and the future

Fred Pearce

Water will define our world in the twenty-first
century. We humans have always built our
homes near water, beside rivers and oases. But
today, we are drying up the great rivers, draining
underground water reserves and changing the
very climate that brings the rain. This most
fundamental resource is ceasing to be where we
want it, when we want it. And in future, its
presence and absence will rock our civilization.

Stand on the huge levees which hold China's
Yellow River in place as it crosses its flood plain
and most times you will see only a trickle of
water. The river's entire flow is taken upstream
for irrigating fields which feed half a billion
people. Go to Indian villages, most of which now
rely on underground water, and the farmers will
tell you how twenty years ago they could lift
water from the well with a bucket. But now,
thanks to rampant overuse, they have to drill
down hundreds of metres to find water, which is
often laced with poisons such as fluoride. Talk to
Syrians and Iranians whose ancient tunnels
tapping water from beneath the mountains have
run dry; or Pakistanis who have abandoned their
parched fields beside the dried-up Indus river.
Visit southern Spain, where the deserts are
invading; or central Asia, where emptied rivers
have all but dried up the Aral Sea. It is not that
the world's water is running out. Nature cleanses
and recycles most of it every ten days or so via
evaporation and rainfall. But we take three times
more water from nature today than we did a
generation ago. Nature can no longer keep up.

And now we face global warming. Usually,
climate change is measured by rising
temperatures, but my bet is that changing
hydrology will prove to be more important. Most
climate models predict that the wet places will
get wetter while the dry places, the places where
water is already a life-and-death resource, will
get much drier. It is already happening. From
the American West to southern and eastern
Africa, rainfall in the continental interiors is
diminishing. Meanwhile, rain storms and
hurricanes around the coasts grow more intense.
The evidence is growing that this trend will
continue. Droughts and floods are, increasingly,
man-made.

Antarctica without its ice

An impression of how Antarctica looks below its ice. In reality, if the
ice were to melt the land mass would rise as a result of being
relieved of its weight.

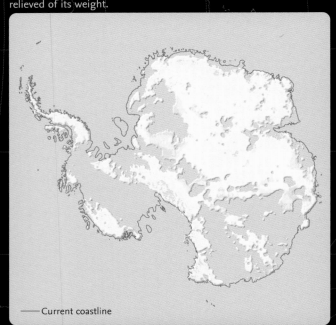

—— Current coastline

And there will be more watery perils coming from the oceans. By mid-century, say glaciologists, ever more of the planet's extra warmth is likely to go into melting ice sheets. The Greenland and West Antarctic ice sheets – which between them hold enough ice to raise sea levels worldwide by 13 m (42.5 feet) – are showing signs of strain. Greenland is losing 1 cubic km (0.2 cubic miles) of ice every forty hours. Jim Hansen of NASA, the US government's top climate modeller, says: 'Once an ice sheet starts to disintegrate, it can reach a tipping point beyond which break-up is explosively rapid'. We could see a rise of 1–5 m (3–16 feet) within the next century, he says – enough to flood hundreds of millions of people from their homes, and render useless the flood defences of great cities, including London's Thames Barrier. We saw the effect of more intense storms, combined with rising sea levels, in New Orleans in 2005. Together, they brought a great city in one of the world's richest and most resourceful countries to its knees. But New Orleans is just the beginning. Imagine a succession of such events in one hurricane season. Imagine similar inundation in Lagos or Bangkok or Sydney or London. Imagine millions of people washed away in Bangladesh or in the Nile delta.

The bottom line is this. Humans have evolved from the Bronze Age to the Broadband Age in about 10 000 years. It has all happened since the end of the last ice age, during a period of environmental and climatic stability. This has been a time when we could plan ahead, build our cities and grow our crops knowing where the water would be and what the weather would do. But, thanks to human interference in the planet's life-support systems, those days are coming to a close. We are probably the last generation to experience the calm before the storm. From now on, the weather will be wild, the seas will rise, and the deserts will advance. Of course, we humans have adaptability in our genes. We would not have survived the ice ages otherwise. But, as events in New Orleans suggested, our modern society may be more vulnerable to nature than we like to think.

What does the future hold?

Michael Allaby

Of all the many types of natural disaster, floods have always been by far the most dangerous. Floods claim more lives and destroy more property than even earthquakes and volcanic eruptions. In years to come the risk from floods may increase in many parts of the world. There are several reasons. The first, and perhaps most significant, is that places vulnerable to flooding are attractive places to live. The flood plains of major rivers are wide, level expanses usually of fertile land, ideal for farming, and also for urban development. At times of very high flow, under natural conditions, low-lying areas absorb excess water which overflows the riverbanks. Flood plain communities must be protected from flooding, however, often by levees. Inevitably there are times, at intervals of decades, when the melting of exceptionally heavy winter snow or prolonged rain deliver more water than a confined river can carry. That is when the levees fail, homes flood, and lives are put at risk. Flash floods may also become more frequent. It is possible, though not certain, that a general rise in temperature will cause mid-latitude weather systems to move more slowly. If that happens, these regions, lying between the tropics and the Arctic and Antarctic Circles, will experience more

prolonged periods of settled weather, but also a greater number of violent storms. Prolonged, even quite moderate, rain can saturate drainage basins, so a single storm can send water cascading down hillsides, carrying with it trees, rocks, and soil. As well as the water, landslides and liquid mud may inundate homes in the valleys.

London in peril?

Since the middle of the twentieth century more and more people have moved to the coast, to live beside the sea. Millions have settled along the eastern seaboard of the United States, on land which is only a few metres above sea level. These homes are at risk from storm surges, when a combination of onshore winds, low air pressure, and an incoming tide carry sea water inland. Low-lying coasts are also at risk from erosion, as the sea steadily washes material away from one place and deposits it in another. Some shorelines advance, but others retreat. The storm surges generated by tropical storms and cyclones are particularly severe. Their frequency changes cyclically, and since the mid-1990s these storms have been growing more frequent. At the same time, higher sea-surface temperatures appear to be making the storms more severe. Maximum wind speeds are not increasing, but more tropical cyclones now attain maximum strength. Not all tropical storms make landfall, but some do. Island and coastal communities are at risk in the Caribbean, the Gulf of Mexico, the southeastern United States, and lands bordering the China Seas.

Prolonged spells of settled weather suggest drought. During a drought clay soils bake hard and crack, but others dry to a granular consistency, and small particles will blow about in the wind. A wind of only 24 km (15 miles) per hour will lift dry sand grains, and if enough sand or dust particles enter rising air there will be a sand or dust storm. Such storms may become more frequent on light soils in mid-latitudes, especially in China, North America, and on a smaller scale possibly southern Europe. Unless measures are implemented to bind the soil, some of these storms could be comparable to the Dust Bowl storms of the 1930s on the North American plains.

Tornadoes may also wreak havoc, not because they will grow more frequent or more severe, but because of a shift in some regions of population from the cities to the countryside. A tornado seldom lasts for longer than a few minutes and most occur in open country. Until fairly recently many tornadoes went unreported, because no one saw them. Rural areas are more densely populated nowadays, which means there is a greater chance that a short-lived tornado will wreck someone's home.

All of these disasters can be avoided, or at least rendered survivable, provided everyone recognizes the risk and prepares for it. In September 2004, a tropical cyclone struck Cuba with winds of almost 260 km (162 miles) per hour, but the authorities were prepared. They evacuated 1.5 million people, together with their pets and most prized possessions. The storm demolished 20 000 homes, but not a single life was lost. It can be done.

The future of this fragile Earth

Guy Dauncey

If you had lived in eighteenth-century Britain or America, you would have lived in a world enmeshed in slavery. If someone had asked, 'What does the future hold for this cruel Earth?' however, your answer would have said more about you than about the future. If you were feeling depressed or defeated about the human condition, you might have replied, 'More of the same'. If you were one of the few who were determined to end the slave trade, your response would have been very different – and you would have been right. Today, we live in a world enmeshed in fossil fuels. The stored ancient sunlight which we release as carbon is playing havoc with the climate, and we are now looking at a 4–6 m (13–20 feet) rise in global sea levels by 2100 due to the increased pace of Greenland's ice-melt. What does the future hold? It is easy to read the science and say, 'More of the same'.

But wait. In Sweden, in October 2005, the government announced a new national policy goal to end Sweden's dependence on fossil fuels by 2020. In Spain, all new housing must be built with a solar hot water system. In Germany, the government's renewable energy policies have led to an enormous growth in the installation of solar, wind, biomass, and other sustainable energy systems. It is impossible to think about the future of our planet without considering the energy we use, and where it comes from. For many thousands of years, our only source of energy aside from the sun came from burning wood. When the firewood began to run out, we started burning coal, a strange black substance which was found in the ground, which we later used to power the industrial revolution and the transformation of the world. For lamps and candles, we killed sperm whales, driving them almost to extinction.

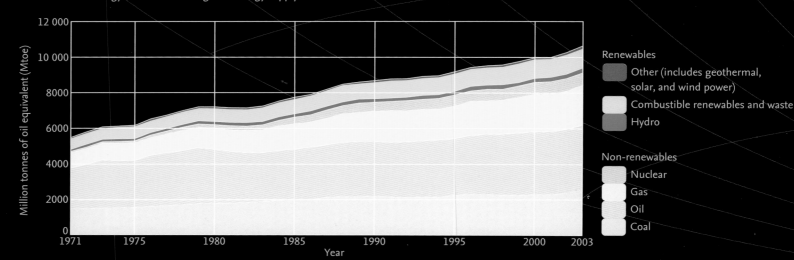

World total energy supply by fuel 1971–2003
Renewable energy in the context of global energy supply

Million tonnes of oil equivalent (Mtoe)

Year

Renewables
- Other (includes geothermal, solar, and wind power)
- Combustible renewables and waste
- Hydro

Non-renewables
- Nuclear
- Gas
- Oil
- Coal

The first oil was drilled in 1848, but the age of oil did not take off until 1901 when huge gushers were drilled in Texas. We did not start using natural gas until the 1890s. The Age of Fossil Fuels will end. Coal, oil, and gas are a one-time benefit from the ancient past, and when they're gone they will be gone forever. We may already be at the peak of the world's oil supply, and as what's left becomes scarce, the price will rise, causing people to seek other sources of energy. The coal will last longer, but until the hope of clean coal with zero carbon emissions is realized, the public pressure to cease burning coal will continue.

When future historians write about this era, they will write about the Age of Fossil Fuels, and how it ended. If we are dumb, their history will tell of conflict, warfare, and chaos, much like the end of the Roman Empire. If we are smart, it will tell of how we achieved an incredible transition to sustainable forms of energy. In spite of what some say, there is an ample supply of solar, wind, and other forms of green, efficient energy which will allow us to continue civilization's quest, if we adjust our lifestyles to live in a sustainable way. If we are smart, the historians will also write of a great transition to organic farming, with accompanying increases in yields, and a transition to green chemistry, ending the growing presence of toxic chemicals in our bodies and in nature. They will also write of a transition to sustainable forms of forest management, ending the ugly looting and pillaging of the world's forests which marked the twentieth century. These things are all within our choice.

During the 1700s, many Quakers pondered how awful slavery was. In 1783, in London, they launched the formal campaign to end it. In 1792 Denmark abolished slavery, followed by France in 1794, Britain in 1834, and America in 1865. Margaret Mead, the great American anthropologist, once said, 'Never doubt that a small group of thoughtful, committed citizens can change the world. Indeed, it is the only thing that ever has'. As humans, we can choose to make a difference. We can choose to love and protect our fragile Earth.

Towards tomorrow

Tim Flannery

We think of the Earth's crust as timeless, unchanging and solid. But in reality the Earth is constantly remaking itself, it's just that we rarely live long enough to see dramatic changes. If you want evidence of past change, just look at the rocks beneath your feet. Depending on where you live they may tell of ancient volcanoes, or perhaps a vanished sea, which occupied the space where you now stand.

The rate at which the Earth changes is a vital factor in determining the fate of living things. Were Earth truly unchanging there would be little life on our planet, and certainly no humanity. That's because soil, ocean nutrients and even our atmosphere eventually wear out and need to be replenished. Our soils, for example, are rejuvenated when volcanoes erupt, mountains erode, or ice sheets grind whole landscapes to dust. Oceans are fed when winds blow dust from deserts, and our atmosphere is added to from volcanic eruptions.

Some parts of the Earth change more quickly than others. Africa's rift valley is a region where volcanoes and mountain ranges form and erode with astonishing rapidity. Little wonder our species, with its large, energy-hungry brain, social habits and omnivorous diet originated there in this land of fertile soils, lakes and biological riches. Indeed there are few other parts of the Earth which could offer similar resources. Australia, on the other hand, is one of the least active continents, and it gives an idea of what a stable Earth might look like. Most of its soils are simply worn out and incapable of supporting anything more than skeletal vegetation and the energy-hoarding marsupials which feed on it.

What happens to living things when change becomes too fast? Common sense tells us that large, fast-moving things can be dangerous. And so it is with changes in climate and topography. Our divisions of the geological time scale are

World population growth 1750–2050

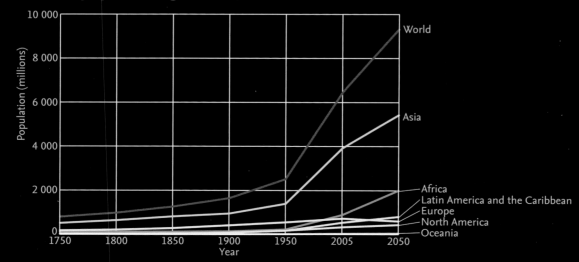

based on extinctions which occurred when conditions on Earth rapidly shifted. Some of these changes were brought about by volcanic eruptions or the impact of asteroids, but most occurred as a result of a changing climate.

Over the past 200 years humans have altered Earth in the most profound ways. In the 1850s tigers were still eating a person per week on the island of Singapore, and the last tiger on the island was not killed until the 1880s. 200 years ago most of the cities of Australia and North America did not exist or were in embryonic form. Back then there was less than a billion people on the Earth, and there were no cars, aeroplanes or electric power plants, and we were burning miniscule amounts of fossil fuel.

Now, in this age of the first global, telekinetic human society, there are six and a half billion of us, and we are drawing down the Earth's reserves of fossil fuels with breathtaking rapidity. And there seem to be cameras everywhere. As shown throughout *Fragile Earth*, cameras on land and in space now allow us to catch nature in the act of change, and even to project how changes will play out in future.

Fragile Earth is an important book because it demonstrates that it is no longer volcanoes or ice ages which are the most important factors changing the face of the Earth. Instead it is us, and in its images we can see how we are treating our only home. Each pair of pictures should prompt us to ask whether we, as individuals and species, are doing the right thing. And if we are not, what we should do about it. These images of change, much of it damaging and resulting from our actions, are just what our species needs to guide it towards a better future.

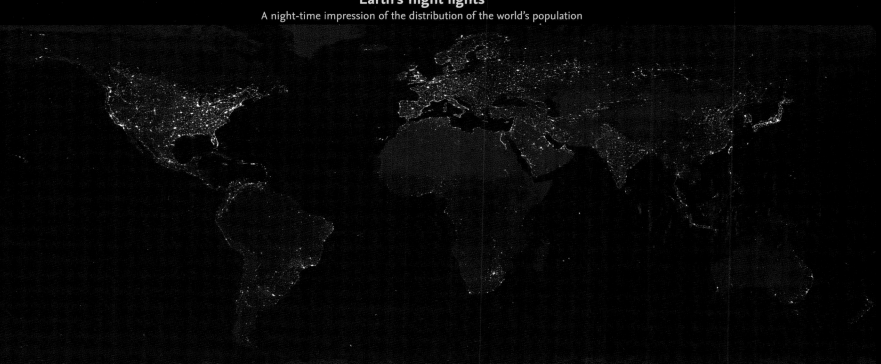

Earth's night lights
A night-time impression of the distribution of the world's population

Observing

Change
Watching the world

arth observation – *looking at the Earth, and monitoring environmental changes, by means of photographs and satellite images*

Satellite orbits

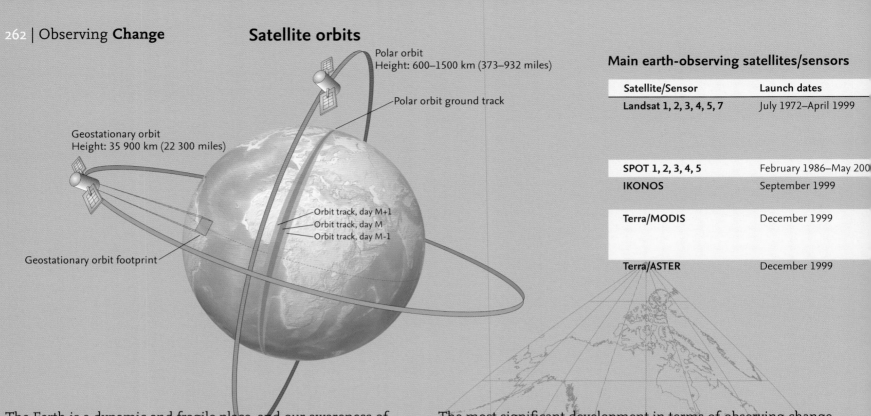

Polar orbit
Height: 600–1500 km (373–932 miles)

Polar orbit ground track

Geostationary orbit
Height: 35 900 km (22 300 miles)

Orbit track, day M+1
Orbit track, day M
Orbit track, day M-1

Geostationary orbit footprint

Main earth-observing satellites/sensors

Satellite/Sensor	Launch dates
Landsat 1, 2, 3, 4, 5, 7	July 1972–April 1999
SPOT 1, 2, 3, 4, 5	February 1986–May 200
IKONOS	September 1999
Terra/MODIS	December 1999
Terra/ASTER	December 1999

The Earth is a dynamic and fragile place, and our awareness of the need to protect it is steadily increasing. In order to direct our efforts, we need to observe and monitor change. We therefore need to gather information on as wide a scale, in as much detail, and as regularly as possible. Traditional means of collecting and recording information – ground survey, scientific measurements, terrestrial photography and mapping – are still as valid today as they have been for centuries. Such methods generate valuable data and dramatic time-sequence images. However, from the earliest days of flight, the ability to observe and photograph the Earth's surface from above has provided perhaps the greatest insight into how the planet looks, works and changes.

The most significant development in terms of observing change from above has been the emergence of Earth-observing satellites. Satellite imagery, and the related science of satellite remote sensing – the acquisition, processing and interpretation of images captured by satellites – is now an invaluable tool in observing and monitoring the Earth at a global level. Sensors carried by satellites can capture data of many types which can be processed to allow detailed interpretation of landscapes, vegetation, and environmental and atmospheric conditions. The level of detail discernible on satellite images is known as resolution. The first Landsat satellites had resolutions of 80 m (262 feet), while the latest satellites now capture images with resolutions of less than 1 m (3.3 feet).

Timeline of remote sensing satellites

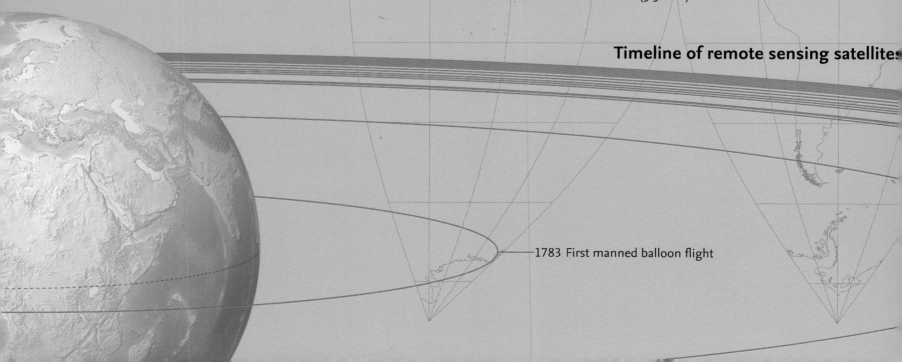

1783 First manned balloon flight

Aims and applications	Resolution	Web address
The first satellite specifically for observing the Earth's surface. Agriculture, geology and numerous environmental and scientific applications.	Landsat 1 and 2: 80 m (Multi Spectral Scanner). Later satellites: 15 m in the panchromatic band (Landsat 7 only), 30 m in the six visible, near and short-wave infrared bands and 60 m in the thermal infrared band.	landsat.gsfc.nasa.gov
Monitoring of land use and water resources, coastal studies and cartography.	Panchromatic 10 m. Multispectral 20 m.	www.spotimage.fr
First commercial high-resolution satellite. Useful for a variety of applications mainly cartography, defence, urban planning, agriculture, forestry and insurance.	Panchromatic 1 m. Multispectral 4 m.	www.geoeye.com
Moderate-resolution Imaging Spectroradiometer. Part of NASA Earth Observing System (EOS). Observation of land, ocean and lower atmospheric processes.	Two bands at nominal resolution of 250 m, 5 bands at 500 m, and 29 bands at 1 km.	modis.gsfc.nasa.gov
Advanced Spaceborne Thermal Emission and Reflection Radiometer. Part of NASA Earth Observing System (EOS). Climatology, hydrology, ecosystem and hazard monitoring.	15–90 m.	asterweb.jpl.nasa.gov

Earth observation

One crucial aspect of remote sensing is that satellites regularly revisit the same point above the Earth and so can gather time-sequence images of exactly the same area. Earth-observing satellites follow one of two types of orbit: geostationary or polar. Geostationary satellites (most commonly meteorological satellites) effectively sit above the same point of the Earth's surface, allowing constant collection of images of the same area. Satellites in polar orbits travel around the Earth in a north-south-north direction. As the Earth rotates they progressively capture images of adjacent (or partially overlapping) areas. Such satellites typically revisit the same point every 16–26 days, but this period can be decreased dramatically if a satellite is able to capture oblique images – the SPOT satellite can in fact revisit the same point every 1–3 days. Such revisit capabilities allow enormous amounts of change data to be gathered.

Satellites and sensors are continuing to be developed and the resolution of the images they capture continues to improve. Sensors recording very detailed environmental and atmospheric data will remain a critical tool in observing the Earth and protecting it from the worst types of change.

1986 SPOT 1
1982 Landsat 4
1978 Landsat 3
1975 GOES-1 First of series of 12 weather satellites; Landsat 2
1972 Landsat 1 First Earth-observing satellite
1990 SPOT 2
1991 ERS-1 European Remote Sensing Satellite
1993 Landsat 6 (failed); SPOT 3
1995 RADARSAT-1 Canadian radar satellite; ERS-2
1997 ORBVIEW-2 First daily colour imagery of the Earth
1998 SPOT 4
1999 Landsat 7; IKONOS First sub-metre resolution commercial satellite; TERRA (NASA Earth Observing System, includes MODIS and ASTER)
1961 First manned space flight
1960 TIROS-1 First weather satellite
1957 First satellite (Sputnik 1)
1903 First powered flight
2000 Shuttle Radar Topography Mission (SRTM)
2001 Quickbird high resolution commercial satellite
2002 SPOT 5; Aqua (NASA Earth Observing System)

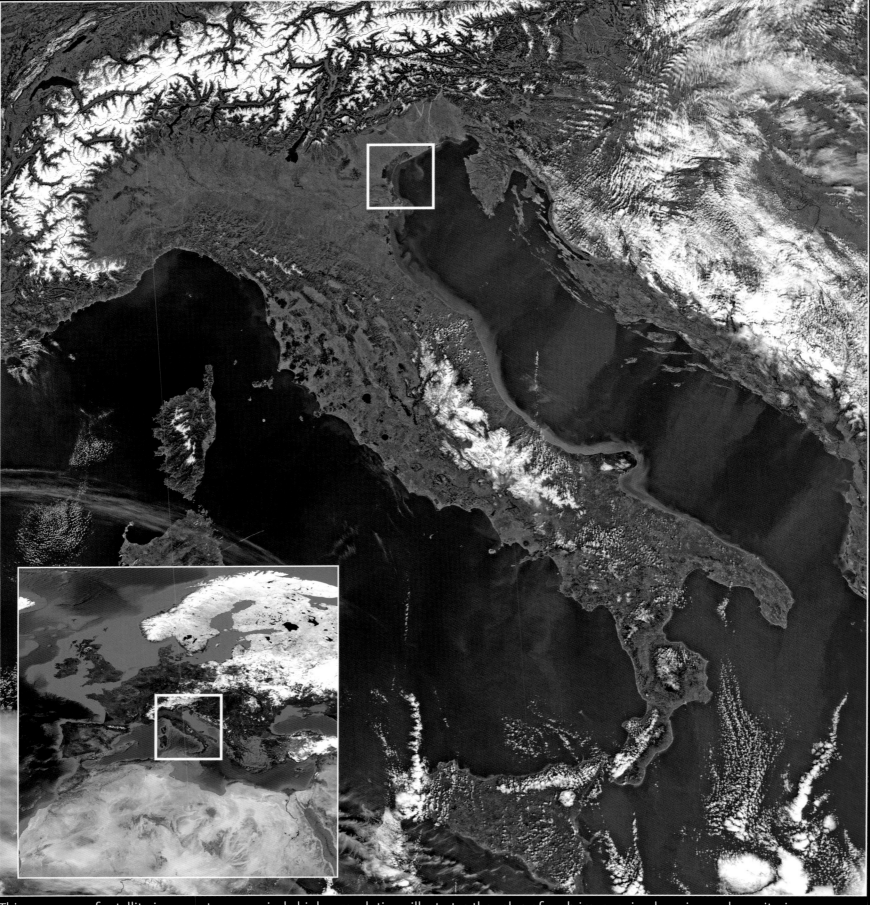

This sequence of satellite images at progressively higher resolutions illustrates the value of such imagery in observing and monitoring features on the Earth. From world, continental and regional views to detailed images picking out individual features on the ground less than 1 m (3.3 feet) in size, images like this can be used for many scientific, environmental and planning purposes.

Satellite sensors commonly collect many types of data which allow very detailed analysis of vegetation and climatic conditions. The fact that Earth-observing satellites regularly revisit the same location adds to this the valuable ability to monitor change over time.

Himalaya, Bhutan

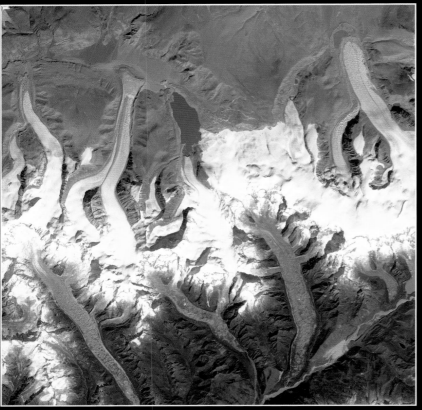

Bourtange, Netherlands

Cape Verde, Atlantic Ocean

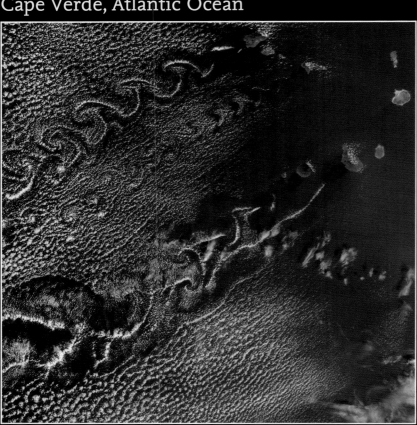

Mississippi Delta, USA

The global reach of Earth-observing satellites enables them to capture images of all aspects of the environment. As well as allowing us to appreciate the beauty of the Earth's varied landscapes, the data collected can be used to measure, monitor and manage change and development. Changes in land use, threats to fragile ecosystems, atmospheric conditions and urban development can all be observed from

Kamchatka Peninsula, Russian Federation

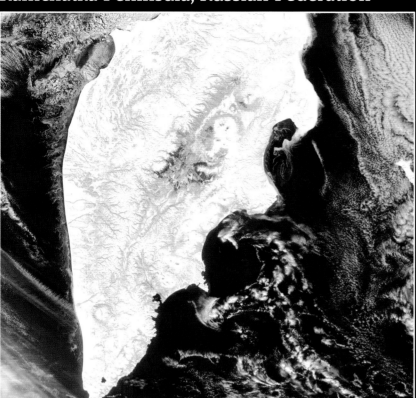

Songhua river, China

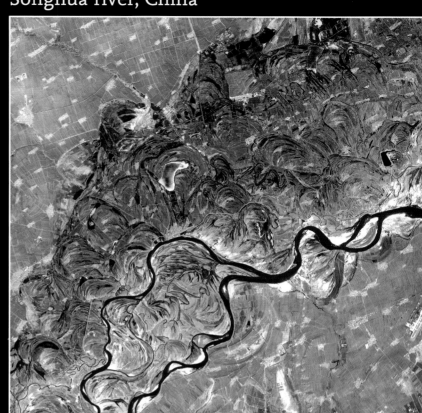

Sydney, Australia

Teshekpuk Lake, Alaska, USA

space and accounted for in planning and in countering damaging activities. Environmental change can be detected sometimes in single images – see the patterns of previous courses of the Songhua river – and, as seen throughout this book, in images of the same locations collected at different times.

Term	Definition
Aquifer	An underground layer of water-bearing, permeable rock from which groundwater can be extracted.
Avalanche	A fall of snow (or rock) down a mountainside when a build up of snow is released down a slope.
Avalanche crown	The top of the avalanche.
Avalanche flank	The side of the avalanche track.
Avalanche trigger	The mechanism or movement which starts an avalanche.
Biodiversity	A measure of the relative diversity among organisms present in different ecosystems.
Biofuel	Fuel which derives from biomass such as recently living organisms or their by-products.
Boreal	Ecosystems local to sub-arctic and sub-antarctic regions.
Carbon cycle	The exchange of carbon between the biosphere, geosphere, hydrosphere and atmosphere of the Earth.
Catchment area	A drainage basin or region of land whose water drains into a river or into a body of water.
Coriolis effect	A force resulting from the rotation of the Earth, affecting winds, ocean currents and atmospheric conditions.
Crown fire	A wildfire which spreads to the top branches of trees sucking oxygen upwards.
Crust	The outer solid layer of a planet. The Earth's crust is between 5 and 60 km (3 and 37 miles) thick.
Cyclone	The rotation of a volume of air around an area of low atmospheric pressure.
Debris toe	Where an avalanche ends when the flow stops.
Deforestation	The conversion of forested areas to non-forested, by the removal of trees usually to create agricultural land.
Desertification	The degradation of land in arid, semi arid and dry sub-humid areas from climate change and human activities.
Drought	An extended period where water availability falls below the requirements for a region.
Earth-observing satellite	A satellite equipped with sensors which observe and record information about the Earth's surface.
Earthquake	A sudden and sometimes catastrophic movement of a part of the Earth's crust.
Ecosystem	The smallest level of organization in nature which incorporates both living and non-living factors.
El Niño	Marked warming of the Pacific Ocean which occurs when an unusually warm current of water moves from the western Pacific, temporarily replacing the cold Peru Current along the west coast of South America.
Evapotranspiration	The return of water to the atmosphere by evaporation from the soil and water bodies and by transpiration from plants.
GDP	Gross Domestic Product. The value of all goods and services produced within a country.
Global warming	An increase in the average temperature of the Earth's atmosphere and oceans which has been observed in recent decades.
Greenhouse effect	The process by which an atmosphere warms a planet.
Greenhouse gas	Gaseous components of the atmosphere which contribute to the greenhouse effect. The major natural greenhouse gases are water vapour, carbon dioxide and ozone.
Ice cap	A dome-shaped ice mass which covers less than 50 000 sq km (19 305 sq miles) of land area.

Term	Definition
Irrigation	Artificial suply of water for agriculture in otherwise fertile areas.
Land reclamation	The creation of new land where there was once water.
Landslide	A sudden movement of a mass of soil or rocks down a cliff or slope.
Levee	A natural or artificial embankment or dyke, usually made of earth, which parallels a river course and protects the adjacent area from flooding.
Mantle	The thick shell of rock surrounding the Earth's outer core, lying between 30 and 2900 km (19 and 1802 miles) below the Earth's crust.
Mudslide	Downward movement of a mass of sediment liquified by rain or by melting snow or ice.
Permafrost	Ground which is permanently frozen.
Pivot-point irrigation	A method of crop irrigation where the equipment rotates around a central pivot giving a circular crop pattern.
Polder	A low-lying tract of man-made land, enclosed by embankments or dykes.
Precipitation	Rain, snow and other forms of water falling to the Earth's surface
Pyroclastic flow	Fast moving bodies of hot gas, ash and rock (collectively known as tephra) emitted from certain types of volcanic eruption.
Sahel	The boundary zone in Africa between the Sahara to the north and the more fertile region to the south.
Sea ice	Ice formed when sea water freezes. Fast ice has frozen along coasts, pack ice is floating consolidated sea ice and an ice floe is a floating mass of sea ice.
Sea level	Mean sea level (MSL) is the average height of the sea, with reference to a suitable reference surface.
Seismic wave	A wave which travels through the Earth, often as the result of an earthquake or explosion.
Stauchwall	The downslope fracture surface of an avalanche.
Storm surge	An onshore rush of water associated with a low pressure weather system, typically a tropical cyclone.
Subduction zone	An area on Earth where two tectonic plates meet, move towards one another, with one sliding underneath the other and moving down into the Earth's mantle.
Tectonic plate	A large piece of the Earth's crust.
Tornado	A violently spinning column of air in contact with both the cloud base and the surface of the Earth.
Tsunami	A series of waves generated when a body of water, such as a lake or ocean is rapidly displaced on a massive scale.
Tundra	Treeless zone between the Arctic ice cap and the northern treeline which has a permanently frozen subsoil (permafrost).
Urbanization	The movement of people from rural to urban areas.
Volcano	A mountain formed from volcanic material ejected from a vent in a central crater.
Water table	The level below which the ground is saturated with water. It rises and falls in response to rainfall and the rate of water extraction.
Wildfire	An uncontrolled fire often occurring in rural or forested areas.

Page numbers in **bold** refer to captions, illustrations and photographs.
Page numbers in *italics* refer to information contained in tables.

272 | Acknowledgements

Contributors

Sir Ranulph Fiennes OBE The world's greatest living explorer and the first person ever to reach both poles by surface travel.

Michael Allaby A prominent environmental science writer who has edited and authored many books and dictionaries on environmental subjects. Author of *A Change in the Weather* and *Facing the Future*.

Guy Dauncey An environmental consultant and freelance writer on environmental, developmental and demographic issues. Author of *Stormy Weather: 101 Solutions to Global Climate Change*.

Tim Flannery Internationally acclaimed scientist, explorer and conservationist. Author of *The Weather Makers: The History and Future Impact of Climate Change*.

Elizabeth Kolbert A reporter for *The New York Times* for fourteen years before becoming a staff writer covering politics for *The New Yorker*. Author of *Field Notes from a Catastrophe*.

Bjørn Lomborg Named one of the 100 globally most influential people by Time magazine in April 2004. Author of *The Skeptical Environmentalist* which challenges widely held beliefs that the global environment is getting progressively worse.

Mark Lynas A broadcast commentator and journalist, and regular writer for *The Guardian, The Observer, The Independent* and *New Statesman*. Author of *High Tide: News from a Warming World* and forthcoming book *Six Degrees*.

Fred Pearce Environmental consultant and leading contributor to *New Scientist*. Author of *When the Rivers Run Dry, Deep Jungle* and *Keepers of the Spring*.

General Acknowledgements

We would like to acknowledge the assistance of :

The United Nations Environment Programme (www.na.unep.net) in providing selected images from *One Planet Many People - Atlas of Our Changing Environment*, UNEP, 2005 (www.earthprint.com).

Page 104 Christopher Riches for kind permission to reproduce an image from *Picturesque Hongkong*. Photos by Denis H. Hazell, published by Ye Olde Printerie Ltd, Hongkong, circa 1920.

Page 147 Clean Air Initiative for Asian Cities (2006) for graph of air quality.

Pages 176–7, 184–5, 230–1 Gary Braasch (worldviewofglobalwarming.org) for images from the forthcoming book *Earth Under Fire: How Global Warming is Changing the World* (University of California Press, 2007).

Pages 156–7 National Snow and Ice Data Center. Boulder, CO. USA for sea ice index images. Digital Media. Fetterer, F. and Knowles, K. 2002 updated 2005.

Pages 182–3 University of Colorado, CO. USA for image of global mean sea level 1993–2005 sealevel.colorado.edu. Leuiette, E. W, Nerem, R.S, Mitchum, G. T, 2004 Calibration of TOPEX/Poseidon and Jason altimeter data to construct a continuous record of mean sea level change. *Marine Geodesy*, 27 (1–2), 79–94.

Page 256 Organisation for Economic Co-operation and Development and International Energy Agency for graph of World Total Energy supply 1971–2003, © OECD/IEA 2005.

Page 259 Data courtesy Marc Imhoff of NASA GSFC and Christopher Elvidge of NOAA NGDC. Image by Craig Mayhew and Robert Simmon, NASA GSFC.

GeoEye www.geoeye.com
National Snow and Ice Data Center. Boulder, CO. USA nsidc.org
NASA earthobservatory.nasa.gov
NASA rapidfire.sci.gsfc.nasa.gov
NASA asterweb.jpl.nasa.gov/index.asp
United States Geological Survey www.usgs.gov

Image credits

16 © Richard Powers/CORBIS
17 HASAN SARBAKHSHIAN/AP/EMPICS
18 Getty Images/ Pictorial Parade/Justin Sullivan
19 Getty Images/Hulton Archive/Justin Sullivan
20 top Pacific Press Service/Alamy
20 bottom Jon Arnold Images/Alamy
21 Pacific Press Service/Alamy
26 top USGS/Cascades Volcano Observatory/Jim Nieland, U.S. Forest Service
26 bottom USGS/Cascades Volcano Observatory/Lyn Topinka
27 top USGS/D.A. Swanson
28 top POPPERFOTO / Alamy
28 bottom © CORBIS
29 top FLPA / Alamy
29 bottom ARCTIC IMAGES / Alamy
30 left © Patrick Robert/Sygma/CORBIS
30 right Andrew Woodley / Alamy
31 BERNHARD EDMAIER / SCIENCE PHOTO LIBRARY
34 IKONOS image © CRISP 2004
35 IKONOS image © CRISP 2004
36 IKONOS image © CRISP 2004
37 IKONOS image © CRISP 2004
38 top © Reuters/CORBIS
38 bottom © DARREN WHITESIDE/Reuters/CORBIS
39 top GREG BAKER/AP/EMPICS
39 bottom © Tomas Van Houtryve/CORBIS
41 © Royalty-Free/CORBIS
42 top PA/PA/EMPICS
42 bottom John Giles/PA/EMPICS
43 Image John Giles/PA/EMPICS
44 top Peter Schneider/AP/EMPICS
44 bottom Hans Rudolf Burgener/Greenpeace
45 Earth Observatory/Space Imaging
45 Earth Observatory/Space Imaging
46 Photo New Zealand/ Nick Groves
47 Photo New Zealand/ Nick Groves
53 MODIS/NASA
54 IKONOS image courtesy of GeoEye
55 top © Vincent Laforet/epa/CORBIS
55 bottom © Mike Theiss/Jim Reed Photography/CORBIS
56 left Courtesy of Alexey Sergeev
56 right © Rick Wilking/Reuters/CORBIS
57 IKONOS image courtesy of GeoEye
58 © Smiley N. Pool/Dallas Morning News/CORBIS
59 © Smiley N. Pool/Dallas Morning News/CORBIS
62 Michel Gunther/Still Pictures
63 Michel Gunther/Still Pictures
64 top © David Hies/CORBIS SYGMA
64 bottom John Russell/AP/EMPICS
65 © Eric Nguyen/Jim Reed Photography/CORBIS
66 IKONOS image courtesy of GeoEye
67 Mike Lawrence/AP/EMPICS
69 MODIS/NASA
70 Sgt Shannon Arledge, USMC/AP/EMPICS
71 top Image State/Alamy
71 bottom Lou Linwei/Alamy
72 NOAA/Department of Commerce
73 NOAA/Department of Commerce
75 MODIS/NASA
76–77 NASA/GSFC
78–79 NASA/GSFC
84 Image reproduced by kind permission of UNEP
85 Image reproduced by kind permission of UNEP
86 IKONOS image courtesy of GeoEye
87 USGS/ASTER
88 top Robert Harding Picture Library Ltd/Alamy
88 bottom Ulana Switucha/Alamy
89 NASA/Japan ASTER Science Team
92 Image reproduced by kind permission of UNEP
93 Image reproduced by kind permission of UNEP
94 Image reproduced by kind permission of UNEP
95 Image reproduced by kind permission of UNEP
96 top © Bettman/CORBIS
96 bottom Lennox McLendon/AP/EMPICS
97 top Peter van Agtmael/Polaris
97 bottom © China Photos/Reuters/CORBIS
98 Image reproduced by kind permission of UNEP
99 Image reproduced by kind permission of UNEP
105 © Setboun/CORBIS
106 © Bettman/CORBIS
107 Paul Russell/CORBIS
108 Image reproduced by kind permission of UNEP
109 Image reproduced by kind permission of UNEP
110 Mark Edwards/Still Pictures
111 Alexandre Meneghini/EMPICS
114 © Bisson Bernard/CORBIS SYGMA
115 guichaoua/Alamy
116 © Crown Copyright/MOD. Reproduced with the permission of the Controller of Her Majesty's Stationery Office
117 GetMapping
118 NASA/Japan ASTER Science Team
119 © Michael Reynolds/epa/CORBIS
126–127 Image reproduced by kind permission of UNEP
128–129 Image reproduced by kind permission of UNEP
130–131 S.Rocha/UNEP / Still Pictures
132 Jacques Jangoux / Still Pictures
133 MARK EDWARDS / Still Pictures
136 AFP/Getty Images
137 © Tom Wagner/CORBIS SABA
138 Image reproduced by kind permission of UNEP
139 left Handout/epa/Corbis
139 right © Francois Carrel/Montagne Magazine/CORBIS
140 © Michel Setboun/CORBIS
141 © Michel Setboun/CORBIS
142–143 INTA/Space Turk
144 INTA/Space Turk
145 © REINHARD KRAUSE/Reuters/CORBIS
148 top RUPERT BUCHELE / WWI / Still Pictures
148 bottom JOHN SHAW/NHPA
149 left © M. Winkel/A.B./zefa/CORBIS
149 right © Ted Spiegel/CORBIS
150–151 © David Woodfall/Woodfall Wild images
158 MODIS/NASA
159 MODIS/NASA
160 MODIS/NASA
161 MODIS/NASA
162 Bryan & Cherry Alexander Photography
163 Tony A. Weyiouanna Sr. Kawerak Transportation Program
164 top Bryan & Cherry Alexander Photography
164 bottom Ashley Cooper/Woodfall Wild Images
165 top Bryan & Cherry Alexander Photography
165 bottom Frank Todd/Bryan & Cherry Alexander Photography
168–169 Greenpeace International
170 Lonnie G. Thompson, The Byrd Polar Research Institute, Ohio State University
171 top Still Pictures/Sidney Paige
171 bottom Still Pictures/Bruce F. Molina
172 top NSIDC/William O. Field
172 bottom NSIDC/Bruce F. Molina
173 NSIDC/Bruce F. Molina
174 Dave Pattison/Alamy
175 © Nina Schwendemann/Reuters/CORBIS
176–177 NASA/Earth Observatory/ASTER data
178 Gary Braasch
179 Gary Braasch
180 National Snow and Ice Data Center, University of Colorado, USA
181 Photo New Zealand/Stewart Nimmo
184 Gary Braasch
185 Gary Braasch
186 IKONOS image courtesy of GeoEye
187 top Shahee Ilyas
187 bottom Jack Sullivan/Alamy
188 US Geological Survey, Center for Coastal Geology
189 US Geological Survey, Center for Coastal Geology
194–195 Voltchev/UNEP/Still Pictures
196 MODIS/NASA
197 top © Jose Fuste Raga/CORBIS
197 bottom © M. ou Me. DESJEUX, Bernard/CORBIS
198 © Nic Bothma/epa/CORBIS
199 Mark Henley/Panos
202–203 Altitude/Still Pictures
204 NASA/Earth Observatory
205 left © Reuters/CORBIS
205 right Mark Edwards/Still Pictures
206 David Woodfall/Woodfall Wild Images
207 top Jeff Henry/Still Pictures
207 bottom Jeff & Alexa Henry/Still Pictures
210 Image reproduced by kind permission of UNEP
211 Image reproduced by kind permission of UNEP
212 1973, 1986, 2001 USGS, EROS Data Center, Sioux Falls, SD
212 2005 MODIS/NASA
213 Gerd Ludwig/Visum/Panos
214 Image State/Alamy
215 © 2006 Shunya
216 Courtesy of John Dohrenwend
217 left Andre Jenny/Alamy
217 right AFP/Getty Images
222 top David Woodfall/Woodfall Wild Images
222 bottom Stacey Peak Media
223 US Geological Survey Center for Coastal Geology
224 © Parks Victoria/Handout/Reuters/CORBIS
225 © Parks Victoria/Handout/Reuters/CORBIS
226 Image reproduced by kind permission of UNEP
227 Image reproduced by kind permission of UNEP
228 Dylan Banarse/Bettina Furnee; www.ifever.org.uk
229 top Simmons Aerofilms
229 bottom Patricia & Angus Macdonald, Aerographica
230 Gary Braasch
231 Gary Braasch
236–237 NASA and U.S./Japan ASTER Science Team
238 NASA
239 NASA/ Earth Observatory/ MODIS data
240 top Paul Glendell/Alamy
240 bottom Steve Sant/Alamy
241 top © Adam Woolfitt/CORBIS
241 bottom Alexander Ruesche/DPA/EMPICS
242 NASA/Earth Observatory/Landsat data
243 NASA/Earth Observatory/Landsat data
254 © Peter Horree/Alamy
264 main NASA/MODIS
264 inset NASA/GSFC
265 main NASA/ASTER
265 inset IKONOS image courtesy of GeoEye
266 top left NASA/ASTER
266 top right IKONOS image courtesy of GeoEye
266 bottom left NASA/MODIS
266 bottom right NASA/ASTER
267 top left NASA/MODIS
267 bottom left IKONOS image courtesy of GeoEye
267 right NASA/ASTER

Collins